# EDEXCEL
# GCSE MATHS
## HIGHER

# HOMEWORK BOOK

Powered by **MyMaths**.co.uk

**OXFORD**
UNIVERSITY PRESS

**OXFORD**
UNIVERSITY PRESS

Great Clarendon Street, Oxford, OX2 6DP, United Kingdom

Oxford University Press is a department of the University of Oxford.
It furthers the University's objective of excellence in research,
scholarship, and education by publishing worldwide. Oxford is a
registered trade mark of Oxford University Press in the UK and in
certain other countries

British Library Cataloguing in Publication Data
Data available

978-0-19-835155-9

10 9 8 7 6 5 4 3 2

Paper used in the production of this book is a natural, recyclable
product made from wood grown in sustainable forests.
The manufacturing process conforms to the environmental
regulations of the country of origin.

Printed in Great Britain

**Acknowledgements**
Although we have made every effort to trace and contact all
copyright holders before publication this has not been possible in all
cases. If notified, the publisher will rectify any errors or omissions at
the earliest opportunity.

Links to third party websites are provided by Oxford in good faith
and for information only. Oxford disclaims any responsibility for
the materials contained in any third party website referenced in
this work.

# Contents

# 1.1 Place value and rounding

**1** Write these sets of numbers in ascending (increasing) order.

  **a** 0.04, 0.14, 0.004, 4, 1.4      **b** 3.92, 9.32, 3.29, 32.9, 0.329

**2** Round these numbers to the nearest

  **i** 10           **ii** 100.

  **a** 493       **b** 207         **c** 94

  **d** 1046      **e** 17 320      **f** 36 575

**3** Round these numbers to the degree of accuracy given in brackets.

  **a** 5376 (nearest 1000)      **b** 67 (nearest 10)

  **c** 63.2 (nearest unit)       **d** 24.265 (2 decimal places)

  **e** 4289 (nearest 100)       **f** 1.96 (1 decimal place)

**4** Round these numbers to 2 significant figures.

  **a** 369       **b** 6555        **c** 0.071 944

  **d** 94.49     **e** 2.955       **f** 15.638

**5** Round these numbers to the degree of accuracy given in brackets.

  **a** 1.347 (3 significant figures)     **b** 12.831 (3 significant figures)

  **c** 0.004 53 (2 significant figures)   **d** 0.3004 (2 significant figures)

  **e** 1239 (2 significant figures)     **f** 63 920 (1 significant figure)

  **g** 235.95 (4 significant figures)   **h** 10 957 (3 significant figures)

**6** Calculate

  **a** $16.89 \times 100$       **b** $439.8 \div 100$

  **c** $6529 \div 1000$       **d** $2.65 \times 1000$

**7**  **a** Alan is 74 and Gena is 66. Alan says they are both the same age when their ages are rounded to 1 significant figure. Is Alan right?

  **b** In how many years' time will Alan be 10 years older than Gena when their ages are rounded to 1 significant figure?

# 1.2 Adding and subtracting

**1** Calculate

   **a** $-8 - 3$       **b** $14 - -6$       **c** $-3 + 10$

   **d** $-9 + 7$       **e** $-3 - 17$       **f** $-10 - -5$

   **g** $-4 + 100$      **h** $17 - -50$      **i** $3 + -5 - -8$

   **j** $19 - -4 + 8$     **k** $-3.2 + -4.5$    **l** $-8.5 - -6.4$

**2** Use a mental method for each of these calculations.
Write the method you have used.

   **a** Increase 35 by 17.

   **b** Decrease 98 by 45.

   **c** What do you need to add to 745 to get 975?

   **d** How many more than 1398 is 2475?

**3** Work out these calculations using a mental method.

   **a** $3.8 + 4.3$      **b** $6.9 + 2.4$      **c** $7.4 - 2.1$

   **d** $5.7 - 2.9$      **e** $32.5 + 7.9$     **f** $45.3 - 5.8$

**4** Work out each of these questions using written methods.

   **a** $6.98 + 4.5$      **b** $12.05 - 3.27$

   **c** $3.52 + 0.596$    **d** $23.5 - 19.41$

**5** Use a written method to work out

   **a** $45.9 + 18.3$     **b** $32.56 - 18.37$

   **c** $82.5 + 9.36$     **d** $12.9 - 4.38$

   **e** $2.08 + 0.589$    **f** $8.03 - 0.214$

**6** Work out each of these questions using written methods.

   **a** $5.64 + 3.2$      **b** $3.5 + 0.482$

   **c** $72.85 + 12.002$   **d** $7.85 - 5.93$

   **e** $24.35 - 18.9$    **f** $15.04 + 2.681$

   **g** $54.862 - 35.97$   **h** $104.25 - 68.489$

**7** Work out $100 - 99 + 98 - 97 + \ldots + 4 - 3 + 2 - 1$

# .3 Multiplying and dividing

**1** Calculate

   **a** $2 \times -3$       **b** $-7 \times -9$       **c** $-12 \div -3$

   **d** $24 \div -6$       **e** $-5 \times 12$       **f** $9 \times -3$

   **g** $-35 \div -7$      **h** $-56 \div 8$       **i** $10 \div -\dfrac{1}{2}$

**2** Work out

   **a** $360 \div 100$     **b** $496 \div 100$     **c** $87 \times 100$

   **d** $3.2 \times 100$      **e** $275 \div 100$     **f** $15.2 \div 100$

   **g** $8 \div 100$        **h** $206 \div 100$     **i** $37.2 \times 100$

**3** Work these out without using a calculator.

   **a** $3 \times 0.2$       **b** $0.05 \times 6$      **c** $12 \div 0.3$

   **d** $21 \div 0.07$     **e** $0.4 \times 0.15$    **f** $4.5 \div 0.9$

   **g** $1.2 \times 0.011$    **h** $2.25 \div 0.015$

**4** Work out these questions without using a calculator.

   **a** $0.4 \times 0.3$     **b** $5 \times 0.8$       **c** $1.2 \div 0.04$

   **d** $750 \div 1.5$     **e** $43.5 \times 0.2$    **f** $0.96 \div 0.016$

   **g** $19.6 \div 0.14$    **h** $1.32 \times 0.3$

**5** Copy these calculations, inserting brackets *if necessary* to make the answers correct.

   **a** $12 \times 3 + 4 = 84$         **b** $24 \div 6 \times 2 = 8$

   **c** $5 \times 4 + 2 \times 7 = 210$    **d** $40 \div 4 + 2^2 = 5$

   **e** $5 \times 4^2 \div 8 = 10$       **f** $15^2 - 10 \times 5 = 175$

   **g** $\dfrac{8^2 \div 2^3 \times 4}{2} = 1$      **h** $\sqrt{150 - 7^2 - 4 \times 5} = 11$

**6** Use a written method to work out

   **a** $5.3 \times 4.6$     **b** $1.57 \times 2.9$     **c** $31.2 \times 5.3$

   **d** $37.9 \times 3.1$     **e** $46.52 \times 19.1$    **f** $5.75 \times 3.17$

**7** Calculate

   **a** $2464 \div 44$     **b** $2162 \div 23$

   **c** $1767 \div 51$     **d** $19\,512 \div 36$

1   Round these numbers to the degree of accuracy given in brackets.

    **a**   4.329 (2 decimal places)

    **b**   3.825 (2 significant figures)

    **c**   0.215 63 (3 decimal places)

    **d**   0.001 85 (1 significant figure)

    **e**   10 214 (3 significant figures)

    **f**   1.200 45 (4 significant figures)

    **g**   1.995 (2 decimal places)

    **h**   45 126 (2 significant figures)

2   Calculate each of these.

    **a**   $6 + -3$                **b**   $-10 \times -6$

    **c**   $-3 \div 100$           **d**   $28 - -9$

    **e**   $-3 \times -8$            **f**   $0.7 \times -100$

    **g**   $51 \div -3$             **h**   $-3 + -12$

3   Do not use a calculator in this question.

    **a**   $297.34 + 109.2$    **b**   $476.92 - 187.48$

    **c**   $65.3 \times 9$           **d**   $27.25 \times 0.7$

    **e**   $102.4 \div 8$          **f**   $254.7 \div 9$

    **g**   $13.16 \div 0.2$       **h**   $9.7 \times 5.4$

4   Use a written method to calculate

    **a**   $561 \times 49$          **b**   $298 \times 27$

    **c**   $496 \div 16$          **d**   $966 \div 23$

5   Evaluate these calculations.

    **a**   $4 + 8 \times 3$       **b**   $12 \div 2 + 4$       **c**   $3 + 1 \times 8 + 4$

    **d**   $5 \times (7 + 4)$     **e**   $3 \times 4^2$          **f**   $3^2 + 7 \times 4$

    **g**   $(8 - 6)^3 \div 4$    **h**   $\dfrac{3 \times (6^2 - 10)}{6}$

# Simplifying expressions

**1** Simplify these expressions.

   **a**   $4x + y + 3x$             **b**   $8x - y + 2y$

   **c**   $x - 3y + 2x$           **d**   $x - 4y + 6z - 5x - 6x + y$

   **e**   $6x + y - 4x - y + 2x$      **f**   $9x - 4y + 2x - 6y$

   **g**   $x + y - x - y$            **h**   $9x + 6y - z - 4y + 2z$

**2** Simplify these expressions.

   **a**   $3a + b + 3a + 2b$        **b**   $5d + 4e - 2d + e + d$

   **c**   $4f + 5g + 6f - 7g + f$     **d**   $4h - i + 3h - 5i + h$

   **e**   $3j + 5k + k - 5k - k$      **f**   $6l + 4m - 5l - 4m - l + m$

   **g**   $q \times 6 \times p$             **h**   $a \times 5 \times b \times a$

   **i**   $b \div 6$                 **j**   $10p \div 5$

**3** Evaluate these expressions when $x = 3$.

   **a**   $2x + 7$        **b**   $5 - x$            **c**   $x^2 + 2x$

   **d**   $\dfrac{3x - 4}{5}$      **e**   $2x^2$            **f**   $\dfrac{2x - 1}{6x - 3}$

**4** Evaluate these, given that $a = -2$.

   **a**   $3a + 8$          **b**   $5a - 1$

   **c**   $10 - a$         **d**   $\dfrac{a}{2}$

   **e**   $a^2 - 3a + 2$      **f**   $\dfrac{2a + 1}{3}$

   **g**   $2a^3$            **h**   $3a^2 - 2a$

**5** Substitute the values $a = 2$, $b = 3$, $c = \dfrac{1}{2}$ into each expression.

   **a**   $4a + b$          **b**   $ac + 1$

   **c**   $2b - a$         **d**   $6b + c$

   **e**   $2a + b + 5c$      **f**   $3c + 2a - b$

   **g**   $(a + b)^2$        **h**   $2a^2 - 2c$

   **i**   $\dfrac{bc}{a}$           **j**   $3ab + 2bc$

**6** If $x = 2$ and $y = -4$, find the value of

   **a**   $3x + 2y$          **b**   $3x - y$

   **c**   $4y - x^3$         **d**   $2y^2 + x^2$

# Indices

**1** Simplify each of these expressions.

**a** $\dfrac{a^3 \times a^3}{a^2}$

**b** $\dfrac{q^5 \times q^2}{q^4}$

**c** $\dfrac{k \times k^3}{k^2}$

**d** $\dfrac{t \times t^3 \times t^2}{t^4}$

**e** $\dfrac{u^4 \times u^3 \times u^2}{u}$

**f** $\dfrac{s \times s^2 \times s^2}{s^3 \times s}$

**g** $\dfrac{p \times p \times p^3}{p^2}$

**h** $\dfrac{r^5 \times r \times r^2}{r^3}$

**i** $\dfrac{v^5 \times v^3 \times v}{v^4}$

**2** Simplify these expressions.

**a** $x^2 \times x^5$

**b** $y^{-3} \div y^4$

**c** $\dfrac{t^2}{t^3}$

**d** $\dfrac{a^3 \times a^5}{a}$

**e** $(b^{-2})^4$

**f** $\dfrac{2x^3 \times 3x^6}{x^4}$

**g** $5p^3 \times 2p^2q$

**h** $(4m^{-2})^3$

**3** Simplify

**a** $4p^2 \times 3$

**b** $4r^2 \times 2s^3$

**c** $4m^3 \times 2m$

**d** $\dfrac{y^2 \times y^3 \times y^4}{y \times y^5}$

**4** Simplify these expressions.

**a** $\dfrac{x^5 \times 3x^2}{x^4}$

**b** $(2m^5)^2$

**5** Work out these, leaving your answer in index form.

**a** $x^2 \times x^4 \times x^8$

**b** $\dfrac{a^5 \times a^9}{(a^2)^3}$

**c** $\dfrac{y}{\left(y^{\frac{1}{2}} \times y^{\frac{1}{2}}\right)^3}$

**6** If $a = 3^2$ and $b = 3^5$ work these out, leaving your answer as an index number.

**a** $ab$

**b** $\dfrac{a}{b}$

**c** $a^2$

**d** $ab^2$

**e** $(3b)^2$

**f** $(ab)^2$

1033, 1045, 1064, 1301 **SEARCH**

# Expanding and factorising 1

**1**  Expand these expressions and simplify where possible.

    **a**  $3(x + 4)$              **b**  $y(y - 3)$

    **c**  $-5(p + 2q - r)$    **d**  $4m(m - n)$

    **e**  $2(a + 3) + 5(b + 4)$    **f**  $2(x + y) - 4(3x - y)$

    **g**  $5(h + 2) - 4(2 - h)$    **h**  $(p - q) - (2 - q)$

**2**  Factorise these expressions.

    **a**  $5x + 10$              **b**  $3x - 9$

    **c**  $12x - 2t$          **d**  $4ab - 2a$

    **e**  $6xy + 9x - 12y^2$     **f**  $8h^3 - h$

**3**  Factorise these expressions.

    **a**  $(a + b) + 3(a + b)^2$    **b**  $(p + qr)^2 - 6(p + qr)$

    **c**  $pq + rq + px + rx$    **d**  $xy + xw - 2y - 2w$

**4**  Factorise these expressions by removing common factors.

    **a**  $6p + 3$            **b**  $12x - 15$

    **c**  $3xy + 2x$        **d**  $4y^2 + 12y^3 + xy$

    **e**  $2pq^2 + 5p^2q$      **f**  $6a^3b - 3a^2 + 12$

    **g**  $2(x + y) - (x + y)^2$    **h**  $wx + wy - 3x - 3y$

**5**  Use factorisation to help you evaluate these without using a calculator.

    **a**  $3 \times 0.43 + 3 \times 1.57$

    **b**  $5 \times 4.93 - 5 \times 2.73$

    **c**  $4.78^2 + 4.78 \times 5.22$

**6**  **a**  Factorise these expressions.

      **i**  $x(x + 1) + 3(x + 1)$    **ii**  $2y(y - 4) - 5(y - 4)$

    **b**  Factorise these expressions.

      **i**  $x^2 + 5x$         **ii**  $2x + 10$

      **iii**  Using your answers to part **b**, factorise the expression $x^2 + 7x + 10$.

# 2.4 Algebraic fractions

**1** Simplify these fractions.

**a** $\dfrac{q^4}{q^8}$     **b** $\dfrac{n^8}{n^3 \times n^3}$

**2** Simplify these expressions, leaving your answer in index form.

**a** $\dfrac{3^5 \times 3^2}{3^4}$     **b** $\dfrac{(5^3)^4}{5^2 \times 5^5}$

**c** $\dfrac{(x^8 \div x^3)^2}{x^4 \times x}$     **d** $y^3 \times \dfrac{y^2 \times y^6}{(y^4)^2}$

**3** Simplify these fractions fully and find the odd one out.

$\boxed{\dfrac{18x}{6}}$   $\boxed{\dfrac{15x^2}{5x}}$   $\boxed{\dfrac{6abx}{2ab}}$   $\boxed{\dfrac{4x - 12x^2}{4x}}$   $\boxed{\dfrac{6x^2 + 3x}{2x + 1}}$

**4** Simplify each of these.

**a** $\dfrac{2a}{5} + \dfrac{a}{5}$     **b** $\dfrac{2x}{3} + \dfrac{x}{6}$

**c** $\dfrac{3}{p} - \dfrac{4}{q}$     **d** $\dfrac{x+3}{4} + \dfrac{x+7}{4}$

**e** $\dfrac{2x+1}{5} - \dfrac{3x+2}{3}$     **f** $\dfrac{4}{a+2} - \dfrac{3}{a-3}$

**5** Simplify these fractional multiplications and divisions.

**a** $\dfrac{2x}{3} \times \dfrac{3}{4}$     **b** $\dfrac{8xy}{9} \times \dfrac{3}{y}$     **c** $\dfrac{x}{5} \div \dfrac{x}{3}$

**6** Simplify these fractions.

**a** $\dfrac{x^2 + 2x}{3x - 3} \times \dfrac{6x - 6}{2x + 4}$

**b** $\dfrac{2x^2 - 6x}{2x^2 + 5x} \div \dfrac{3x - 9}{12x + 30}$

**c** $\dfrac{y^2 - 3y}{3y + 6} \times \dfrac{4y + 8}{(y - 3)^2}$

Q 1149, 1151, 1164   **SEARCH**

**1** Evaluate these expressions when $a = 2$, $b = -3$ and $c = 0.5$.

  **a**   $2ab$             **b**   $3bc$

  **c**   $\dfrac{b+a}{c}$          **d**   $\dfrac{4ac}{b^2}$

  **e**   $5b(c + a)$        **f**   $\dfrac{6c - 2a}{5 - b}$

**2** Simplify these expressions where possible.

  **a**   $2 \times x$            **b**   $y \times y \times y$

  **c**   $4x + 9x$         **d**   $5a + 3b$

  **e**   $7m - n - 3m - 2n$   **f**   $x^2 + 5x + 3x^2$

  **g**   $3pq + 8qp$       **h**   $2p \times 4q$

  **i**   $\dfrac{9a}{a}$             **j**   $\dfrac{30x^2}{6xy}$

**3** Simplify these expressions where possible.

  **a**   $4x \times 3y$        **b**   $2p \times 4p$

  **c**   $a \times a \times 3a$     **d**   $5ba + 3ab + bc$

  **e**   $3m - n + 4n + 2m$   **f**   $x^2 + 3x^3 - 2x^2$

  **g**   $\dfrac{6gk}{3k^2}$          **h**   $\dfrac{8y^4}{y^3}$

**4**   **a**   Simplify these expressions.

       **i**   $6p + 2p - 3q$

       **ii**   $5a \times 3b$

  **b**   Expand $4(5 - 3x)$.

  **c**   Expand and simplify $4(2y + 1) - 3(5y - 2)$.

**5** Factorise each expression.

  **a**   $4ab - 2a$        **b**   $10x^2 - 15x$

  **c**   $a^3b^2c + 3a^2c$    **d**   $3p^2 + 9p^3 + pq$

# 3.1 Angles and lines

**1** Calculate the size of the angles marked by letters.

**a**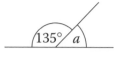

**b**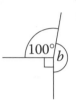

**c**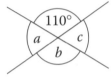

**d**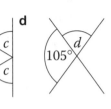

**2** Work out the missing angles, giving reasons for your answers.

**a**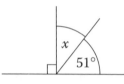

**b**

**c**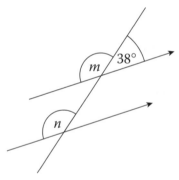

**d**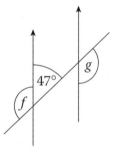

**3** Use the diagram to measure the bearing of

    **a** Coventry from Birmingham

    **b** Worcester from Birmingham

    **c** Birmingham from Coventry

    **d** Worcester from Coventry.

N
Birmingham  Coventry
• Worcester

**4** **a** Draw a diagram to show the position of the points $A$ and $B$ where the bearing of $B$ from $A$ is $112°$.

    **b** What is the bearing of $A$ from $B$?

Q 1082, 1086, 1109    SEARCH

# 3.2  Triangles and quadrilaterals

1  Calculate the third angle of each of these triangles and state the type of triangle.

    **a**  35°, 100°       **b**  45°, 90°       **c**  15°, 120°

    **d**  60°, 60°       **e**  27°, 143°      **f**  75°, 30°

2  Work out the missing angles, giving reasons for your answers.

**a**
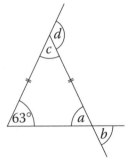

**b**
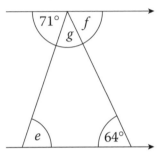

3  Plot the points (1, 3) and (3, 1) on a copy of this grid.

These points are two vertices (corners) of a shape. Using a different grid for each shape, add other vertices to make a

    **a**  square

    **b**  parallelogram

    **c**  trapezium

    **d**  kite.

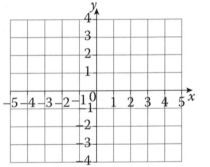

4  Work out the size of the angles marked by letters. Give a reason for each answer.

**a**
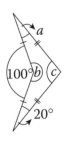

**b**
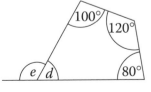

**c**
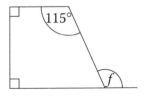

# 3.3　Congruence and similarity

**1**　Which of these pairs of triangles are congruent? Give reasons for your answers.

**A**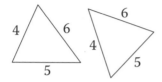

**B**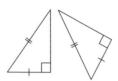

**C**

**D**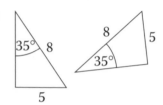

**2**　Betty brings back two boxes of 'Mackay' chocolates from Scotland. Each box has the same picture on the front.

The picture on the 320 g box is 102 mm by 170 mm.

The picture on the 480 g box is 140 mm by 226 mm.

Show that the two rectangles are *not* mathematically similar.

**3**　This diagram shows a triangle *PST*.

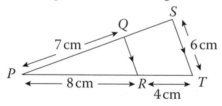

Work out the length of　　**a**　*QR*　　**b**　*PS*.

**4**　Prove that triangle *PQR* is congruent to triangle *RST*.

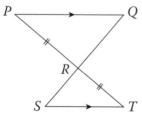

Q 1119, 1148　**SEARCH**

# Polygon angles

**1** Calculate the sum of the interior angles of a

    **a**   pentagon         **b**   hexagon

    **c**   nonagon         **d**   decagon.

> Hint: Split the shapes into triangles.

**2** A regular polygon has 20 sides.

    **a**   Calculate the size of an exterior angle.

    **b**   Calculate the size of an interior angle.

**3** Calculate the size of the unknown angles in these polygons.

    **a**

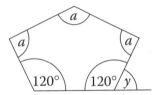

    **b**

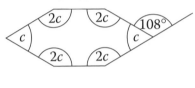

**4** The interior angle of a regular polygon is 120°.

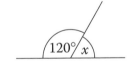

    **a**   Calculate the size of the exterior angle $x$.

    **b**   Calculate the number of sides of the regular polygon.

    **c**   What is the mathematical name for this polygon?

**5**   **a**   The exterior angle of a regular polygon is 30°.
        Write the interior angle of this polygon.

    **b**   The exterior angle of a regular polygon is 40°.
        Work out the number of sides of this polygon.

        Name this polygon.

**1** Find the value of the angles marked by letters.

**a**

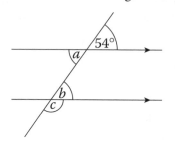

**b**

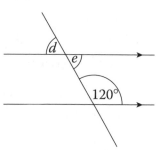

**2** *Y* is due north of *X*. The bearing of *Z* from *Y* is 124°. The bearing of *X* from *Z* is 233°.

   **a** Draw a sketch diagram to represent this information.

   **b** Work out the three angles of the triangle *XYZ*.

   **c** What is the bearing of *Z* from *X*?

**3** *ABCD* is a square. Prove that triangles *ABC* and *ACD* are congruent.

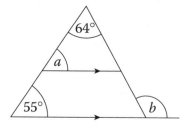

**4** Sketch these shapes. Label equal angles, equal sides and parallel lines.

   **a** rectangle            **b** parallelogram

   **c** equilateral triangle    **d** isosceles triangle

   **e** kite               **f** trapezium

**5** Work out the missing angles, giving reasons for your answers.

**a**

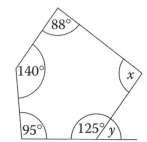

**b**

# Sampling

**1** For each survey, suggest reasons why the chosen sample may be biased.

   **a** To find out the most popular sport amongst young people by asking teenage boys as they leave a football training academy.

   **b** To find out the most popular make of car by noting down the cars belonging to people who live on one road in Nottingham.

   **c** To find out the popularity of school uniforms by telephoning all the households on one page of the telephone directory on a Monday.

   **d** To find out the preferred choice of music for a school disco by asking the students who leave one of these discos at the end of the night.

**2** A machine in a factory produces bottles with screw caps. Every 20th bottle is produced with a cap that does not seal the bottle. A factory worker notices one faulty bottle cap and decides to take a systematic sample of bottles in order to assess the scale of the problem. Explain the effect of sampling

   **a** every 20th bottle, beginning at bottle number 20

   **b** every 5th bottle, beginning at bottle number 5

   **c** every 5th bottle, beginning at bottle number 3.

> Hint: Work out the proportion of faulty bottles in the sample taken and decide on the most likely conclusion of the factory worker.

**3** Charlotte carries out a survey to find out whether or not people belong to a gym. She asks all the people at her hockey club.

   **a** Write down two reasons why this is not a good way to find out whether or not people belong to a gym.

   **b** Devise a question that Charlotte could ask in order to find out how often people use a gym.

**4** The populations of three small towns are 12 400; 15 500 and 17 100. A journalist wants to estimate the likely result of the impending local election, so she telephones 100 residents from each town and asks them their opinion.

   **a** Give two reasons why this is not a good method of sampling the population.

   **b** Find the number of residents she would select from each town if she chooses a stratified sample.

# 4.2 Organising data

**1** These results show how many TVs are owned by students' families.

| Number of TVs | 0 | 1 | 2 | 3 | 4 | 5 |
|---|---|---|---|---|---|---|
| Number of students | 1 | 7 | 12 | 13 | 5 | 1 |

    **a** How many students were included in the survey?

    **b** Calculate the total number of TVs owned by all of the students.

**2** The heights of students in a class, in metres, to the nearest cm, are

1.60   1.45   1.51   1.63   1.70   1.46   1.38   1.44
1.52   1.39   1.50   1.48   1.60   1.52   1.36   1.70
1.63   1.55   1.49   1.36   1.45   1.42   1.51   1.67

Draw an ordered stem-and-leaf diagram to represent this information. Choose suitable stems.

**3** Lynda recorded the Mathematics test results of 30 students.

38   89   52   42   40   35   62   51   54   53
77   71   84   55   65   62   80   74   70   43
31   48   56   66   64   72   70   79   87   82

Represent these data on a stem-and-leaf diagram.

**4** A group of 36 students can choose either French or Spanish as an option. Copy and complete the two-way table.

| | French | Spanish | Total |
|---|---|---|---|
| Boys | 8 | | |
| Girls | | 11 | |
| Total | 16 | | |

Q 1214, 1193 SEARCH

# .3 Representing data

**1** 60 people were asked to name their favourite colour.

The results are shown in the table.

Draw a bar chart to represent this information.

| Colour | Frequency |
|--------|-----------|
| Red | 15 |
| Blue | 12 |
| Green | 18 |
| Purple | 5 |
| Black | 2 |
| Other | 8 |

**2** In a city, 1800 cars were stolen in a year. The table below shows the times of the day when they were stolen.

| Time | Number of cars |
|------|----------------|
| Midnight to 6 a.m. | 700 |
| 6 a.m. to midday | 80 |
| Midday to 6 p.m. | 280 |
| 6 p.m. to midnight | 470 |
| Time unknown | 270 |

Draw a pie chart to show this information.

**3** A fruit seller sells 180 pieces of fruit in a day, as shown in the pie chart.

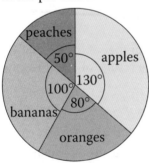

**a** Calculate the angle that represents one piece of fruit.

**b** Find the number of
  **i** bananas **ii** peaches sold.

# 4.4   Averages and spread 1

1   This is the data Priya collected on the amount of pocket money, in pounds, that she received over 15 weeks.

10   8   10   6   10   12   5   8   9   50   14   15   5   10   8

   **a**   Find, for this set of numbers
     **i**   the mean     **ii**   the mode     **iii**   the median
     **iv**   the range     **v**   the interquartile range.
   **b**   What effect did the £50 have on each of your answers in part **a**?

2   Two dice are thrown and their scores added together. The results of 100 throws are shown in the table.

| Score | 2 | 3 | 4 | 5 | 6 | 7 | 8 | 9 | 10 | 11 | 12 |
|---|---|---|---|---|---|---|---|---|---|---|---|
| Frequency | 4 | 4 | 7 | 13 | 10 | 20 | 16 | 12 | 5 | 7 | 2 |

   **a**   Work out the mean score per throw.
   **b**   Work out the
     **i**   median             **ii**   interquartile range.

3   The table below shows the number of letters in the first 50 words of the novel *The Hobbit* by J. R. R. Tolkien.

| Word length | 1 | 2 | 3 | 4 | 5 | 6 | 7 |
|---|---|---|---|---|---|---|---|
| Frequency | 5 | 11 | 12 | 10 | 7 | 4 | 1 |

   **a**   Find the   **i**   mean   **ii**   median   **iii**   mode.
   **b**   Calculate the range.

4   A student throws two dice and the smaller number of the two shown on the dice is subtracted from the larger. The results of 100 throws are shown in the table.

| Result | 0 | 1 | 2 | 3 | 4 | 5 |
|---|---|---|---|---|---|---|
| Frequency | 18 | 26 | 21 | 14 | 14 | 7 |

   **a**   Calculate the median.     **b**   Calculate the interquartile range.

5   These are the warmest January temperatures (in °C) for the decade 1920–29 as measured in Wick and Southampton.

Wick:   5.8   7.2   4.9   7.1   6.1   6.6   6.3   6.2   5.7   4.8

S'ton:   8.6   10.2   8.0   8.9   8.4   9.1   8.1   8.2   9.3   4.4

   **a**   Show the data on a back-to-back stem-and-leaf diagram.
   **b**   Write two comparisons between the data sets.

1   For each of these surveys, suggest reasons why the chosen sample may be biased.

   a   To find out the most popular holiday destination by asking people as they leave a travel agency.

   b   To find out the most popular brand of washing powder by asking people as they go into a supermarket at 11 a.m. on a Tuesday.

   c   To find out how people travel to school by telephoning all the households on one page of the telephone directory on a Wednesday.

2   Doris did a survey of the number of occupants of all the cars passing her house during one hour. Her results are shown in the table.

| No. of occupants | 1 | 2 | 3 | 4 |
|---|---|---|---|---|
| Frequency | 17 | 13 | 9 | 11 |

   a   Work out the mean number of occupants per car.

   b   What is the modal number of occupants?

3   A student tosses five coins and the number of heads is recorded each time. The results of 200 tosses are shown in the table below.

| Number of heads | 0 | 1 | 2 | 3 | 4 | 5 |
|---|---|---|---|---|---|---|
| Frequency | 8 | 30 | 58 | 62 | 34 | 8 |

   a   Calculate the median.

   b   Calculate the interquartile range.

4   These are the mean temperatures (in °C) for March recorded in England and Wales during the last two decades of the 20th century.

   1980 to 1989:  4.3  7.4  5.6  6.0  4.3  4.3  4.6  3.8  5.9  7.0

   1990 to 1999:  7.9  7.4  6.9  6.1  7.1  5.0  4.0  7.8  7.5  6.9

   a   Represent these data on a back-to-back stem-and-leaf diagram.

   b   Compare the two sets of data. You may like to mention the median, mode, range or interquartile range.

1 Calculate these fractions of a quantity. Give your answers as mixed numbers where appropriate.

   a $\frac{1}{4}$ of 48 m          b $\frac{5}{12}$ of 132 kg

   c $\frac{5}{8}$ of 75 km         d $\frac{4}{9}$ of 66 litres

2 Calculate these fractions of a quantity, giving your answers as fractions or mixed numbers where appropriate and showing any cancelling.

   a $\frac{1}{3}$ of 105 m         b $\frac{1}{7}$ of 252 g         c $\frac{3}{4}$ of 52 cm

   d $\frac{2}{5}$ of 55 litres     e $\frac{5}{8}$ of 96 kg         f $\frac{7}{12}$ of 126 mm

   g $\frac{4}{9}$ of 21 km         h $\frac{4}{15}$ of 63 cl

3 Calculate these percentages using an appropriate written method.

   a 5% of £540          b 15% of $620

   c 12% of €625         d 24% of £325

   e 52% of £275         f 73% of $1400

   g 64% of €350         h 87% of £2600

4 Calculate these percentages, giving your answers to 3 significant figures.

   a 52% of £85          b 86% of £127

   c 12% of £46          d 7.5% of £38

   e 37.5% of £72        f 14% of £57

5 Write these percentages as equivalent decimals.
   a 33%      b 75%      c 6.8%      d 0.42%

6 True or false?
To work out 109% of a number, multiply by 1.09.
Show your working.

   Hint: Use fractions.

7 True or false?
25% of $\frac{4}{5}$ of 960 is greater than 45% of $\frac{2}{5}$ of 960.

1030, 1031, 1046 SEARCH

# Calculations with fractions

Do not use a calculator for these questions.

**1** Calculate these additions and subtractions.

a $\dfrac{2}{7} + \dfrac{3}{7}$  b $\dfrac{5}{9} - \dfrac{2}{9}$  c $\dfrac{1}{2} + \dfrac{1}{3}$  d $\dfrac{7}{8} - \dfrac{3}{4}$

e $\dfrac{3}{5} + \dfrac{1}{4}$  f $1\dfrac{1}{3} - \dfrac{5}{12}$  g $2\dfrac{3}{4} + 1\dfrac{5}{6}$  h $4\dfrac{4}{9} - 2\dfrac{1}{6}$

**2** Calculate these multiplications.

a $\dfrac{5}{6} \times \dfrac{1}{8}$  b $\dfrac{4}{5} \times \dfrac{5}{8}$  c $\dfrac{2}{7} \times \dfrac{3}{9}$  d $\dfrac{5}{13} \times \dfrac{3}{5}$

**3** Calculate

a $1\dfrac{1}{4} \times 2\dfrac{2}{3}$  b $2\dfrac{1}{2} \div 1\dfrac{1}{3}$  c $5\dfrac{3}{7} \times 3\dfrac{1}{2}$  d $2\dfrac{5}{8} \div 1\dfrac{3}{4}$

**4** Calculate

a $\dfrac{2}{9} + \dfrac{2}{3}$  b $\dfrac{3}{8} + \dfrac{2}{7}$  c $\dfrac{5}{6} - \dfrac{5}{8}$  d $\dfrac{1}{2} \times \dfrac{4}{9}$

e $\dfrac{5}{12} \times \dfrac{3}{10}$  f $\dfrac{5}{8} \div \dfrac{1}{4}$  g $\dfrac{7}{9} \div \dfrac{2}{3}$  h $3\dfrac{4}{5} - 2\dfrac{1}{3}$

i $4\dfrac{1}{2} \times 2\dfrac{4}{9}$  j $4\dfrac{1}{6} \div 2\dfrac{3}{10}$

**5** Davina spends $\dfrac{1}{8}$ of her pocket money on chocolate, $\dfrac{2}{5}$ on a day out with friends and uses the rest to buy her friend Karen a birthday present. What fraction of her money does Davina spend on Karen?

**6** A charity sells plastic bracelets advertising their cause for £2. Of each bracelet sold, $\dfrac{3}{10}$ of the money is used to pay overheads and $\dfrac{1}{8}$ is used for marketing. The rest of the money is used directly for charity work.
 a How much money from each bracelet is used to pay overheads?
 b How much money is used directly for charity work?

**7** a A college has 1503 students, of which $\dfrac{5}{9}$ are on Modern Apprenticeship courses. How many students are not on Modern Apprenticeship courses?
 b How many 75 cl wine bottles can be filled from a container holding 6.5 litres of wine?
 c In a sale, the price of a mobile phone is reduced by $\dfrac{2}{5}$. The original price was £259.95. What is the sale price?

# 5.3 Fractions, decimals and percentages

**1** Write these fractions as percentages.

    **a** $\dfrac{17}{25}$     **b** $\dfrac{19}{20}$     **c** $\dfrac{113}{200}$     **d** $\dfrac{7}{15}$

**2** Write each of these fractions as hundredths. Then convert each fraction to a decimal.

    **a** $\dfrac{1}{4}$     **b** $\dfrac{4}{5}$     **c** $\dfrac{9}{20}$     **d** $\dfrac{8}{25}$

**3** Use short division to convert each of these fractions to decimals.

    **a** $\dfrac{5}{8}$     **b** $\dfrac{1}{6}$     **c** $\dfrac{1}{16}$

    **d** $\dfrac{8}{9}$     **e** $\dfrac{5}{12}$

**4** Convert 55% to

    **a** a decimal number     **b** a fraction in its simplest terms.

**5** Convert these decimals to fractions.
Give your answers in their simplest form.

    **a** 0.36     **b** 0.85     **c** 0.432

**6** Write each set of fractions in ascending order.

    **a** $\dfrac{7}{10}, \dfrac{4}{5}, \dfrac{3}{4}, \dfrac{9}{20}$     **b** $\dfrac{24}{35}, \dfrac{5}{7}, \dfrac{13}{35}, \dfrac{4}{5}$

**7** Arrange these fractions, decimals and percentages in ascending order.

    **a** $\dfrac{3}{5}, \dfrac{1}{4}$, 32%, 0.57, 0.2     **b** 15%, $\dfrac{1}{5}$, 0.152, $\dfrac{2}{9}, \dfrac{4}{11}$

**8** Write these fractions, decimals and percentages in descending order.

    **a** 50%, 0.42, 0.07, $\dfrac{7}{20}, \dfrac{2}{5}$, 12%, 0.9, $\dfrac{71}{100}$

    **b** 0.7, 28%, 0.09, $\dfrac{1}{25}$, 37%, 1, 82%, $\dfrac{11}{20}$

**9** Write $0.3\dot{6}$ as a fraction. Give your answer in its simplest form.

**10** Use your calculator to find the percentage difference between the actual value of $\pi$ and the number 3.142.

1015, 1016, 1063, 1066   SEARCH

**1** Calculate these fractions of a quantity using mental methods.

    **a**   $\frac{1}{4}$ of 372        **b**   $\frac{3}{5}$ of 1125        **c**   $\frac{3}{7}$ of 5915.

**2** Calculate these percentages.

    **a**   10% of 90 km      **b**   28% of 170 litres      **c**   15% of 70 cm

    **d**   7% of 42 kg        **e**   8% of 60 kg          **f**   34% of 60 litres.

**3** Calculate these, giving your answer in its simplest form.

    **a**   $\frac{1}{4}+\frac{5}{8}$        **b**   $\frac{5}{7}-\frac{2}{5}$        **c**   $3\frac{2}{7}-\frac{5}{8}$        **d**   $\frac{23}{16}-\frac{-8}{15}$

    **e**   $\frac{4}{9}\times\frac{3}{4}$        **f**   $2\frac{1}{2}\div\frac{5}{9}$        **g**   $4\frac{1}{9}\times1\frac{13}{27}$        **h**   $4\frac{2}{3}\div\frac{7}{8}$

**4** Write each set of fractions in ascending order.

    **a**   $\frac{1}{4},\frac{5}{6},\frac{3}{8},\frac{2}{3}$      **b**   $\frac{3}{7},\frac{1}{2},\frac{5}{9}$      **c**   $\frac{3}{5},\frac{7}{15},\frac{18}{30},\frac{9}{10}$

**5** Use short division to convert each of these fractions to decimals.

    **a**   $\frac{3}{8}$        **b**   $\frac{4}{9}$        **c**   $\frac{5}{16}$        **d**   $\frac{7}{12}$

**6** Write these fractions, decimals and percentages in descending order.

    **a**   25%, 0.68, $\frac{24}{100}$, $\frac{9}{20}$, 0.28, 64% $\frac{7}{50}$, 0.56

    **b**   $\frac{3}{20}$, 6%, $\frac{1}{5}$, 29%, 0.07, 50%, 0.3, $\frac{3}{25}$

**7** Work these out using mental methods.

    **a**   $\frac{1}{8}$ of 192      **b**   $\frac{3}{8}$ of 256      **c**   $\frac{1}{2}\times\frac{1}{4}$

    **d**   $\frac{4}{7}+1\frac{1}{3}$      **e**   $\frac{4}{9}\div\frac{2}{3}$

**8** Convert $0.\dot{9}$ to a fraction.

    What integer is equivalent to this fraction?

# 6.1 Formulae

**1** ABC taxis charge £2.50 to hire a taxi and then £1.50 for each mile travelled.

    **a** Write a formula for the cost, £C, of hiring a taxi to travel $d$ miles.

    **b** Use your formula to find

        **i** the cost of a 15 mile journey

        **ii** the distance travelled if the cost was £15.25.

**2** Rearrange each of these formulae to make $x$ the subject.

    **a** $x - a = b$        **b** $px = q$        **c** $A = xy + r$

    **d** $tx - v^2 = w^2$     **e** $P = x(q + r)$     **f** $ax^2 = b$

    **g** $h = c(x - d)$      **h** $x^3 p = qr$       **i** $px - qx = n^2$

**3** Make $x$ the subject of these formulae.

    **a** $x + ab = c$       **b** $p^3 x - q = r$       **c** $d + \dfrac{x}{c} = f$

    **d** $\sqrt{x + k} = 4$      **e** $\sqrt{(n^2 + x^2)} = m$

**4** Change the subject of these equations to that given in brackets.

    **a** $y = 3x + 8$         $(x)$        **b** $s = \dfrac{2t}{3} - r$     $(t)$

    **c** $pq - r = s$        $(q)$        **d** $a = \dfrac{b}{c}$         $(c)$

    **e** $2p - q = t$        $(q)$        **f** $\dfrac{a(b - c)}{x} = y$   $(a)$

    **g** $p(q - r) = a(q + b)$  $(q)$      **h** $a(x - b) = \dfrac{x}{c}$   $(x)$

**5** $A$ appears twice in each of these formulae. Collect the terms in $A$ on one side and rearrange to make $A$ the subject of the formula.

    **a** $aA + b = cA + d$           **b** $Ax + 6 = 2 - Ay$

    **c** $p(A + a) = q(b - A)$      **d** $\dfrac{3 + A}{3 - A} = x$

    **e** $A + x = \dfrac{2A + 3}{x}$         **f** $\sqrt{\dfrac{A - p}{A - q}} = \dfrac{1}{2}$

**6** Use the formula $r = \sqrt{\dfrac{A}{\pi}}$ to find the radius of a circle with area $64\,\text{cm}^2$. Give your answer to 3 significant figures.

Q 1170, 1171, 1186    **SEARCH**

**1** Find the functions shown in these mapping diagrams.

**a**

4
6    $\xrightarrow{\text{f}(x)}$    -8
7                        -12
                         -14

**b**

2
7    $\xrightarrow{\text{f}(x)}$    -4
10                       11
                         20

**2** For each function, find

**i** $f(1)$     **ii** $f(2)$     **iii** $f(3)$     **iv** $f(4)$     **v** $f(5)$

**a** $f(x) = x^2 + 1$

**b** $f(x) = \dfrac{1}{x^2}$

**c** $f(x) = (x + 1)(x - 1)$

**d** $f(x) = \dfrac{x+1}{x}$

**3** Find the inverse of each of the following functions.

**a** $f(x) = 3x + 2$     **b** $f(x) = \sqrt{x+5}$     **c** $f(x) = \dfrac{1}{x-3}$

**4** **a** Plot the graph of the function $f(x) = 2x - 1$ and its inverse on the same grid, using values of $x$ from 0 to 5.

**b** What line on the graph is the mirror line of the two functions you have drawn?

**c** Draw this mirror line on your grid.

**5** $f(x) = 2x - 8$ and $g(x) = \dfrac{1}{x+4}$

Find

**a** $f(0)$     **b** $g(0)$     **c** $f(1)$

**d** $g(1)$     **e** $f(-2)$     **f** $g(-2)$

**g** $f\left(\dfrac{1}{2}\right)$     **h** $g\left(\dfrac{1}{2}\right)$     **i** $f\left(\dfrac{-2}{3}\right)$

**j** $g\left(\dfrac{-2}{3}\right)$     **k** $f(x^2)$     **l** $g(x^2)$

**m** $fg(x)$     **n** $gf(x)$     **o** $fg(1)$

**p** $gf(1)$     **q** $fg(-2)$     **r** $gf(-2)$

# 6.3 Equivalences in algebra

**1** Copy and complete the table. Choose from the words *identity, equation* or *formula* for the right-hand column. The first one has been completed for you.

| | Identity, equation or formula? |
|---|---|
| $a \times a \times a \equiv a^3$ | *identity* |
| $3x + 4 = 9x - 2$ | |
| $4p^2(p-1) = 4p^3 - 4p^2$ | |
| $A = \pi r^2$ | |
| $x^2 = 49$ | |
| $V = lwh$ | |

**2 a** Prove that the product of an odd number and an even number is always even.

**b** Prove that squaring an even number will give a number that is divisible by 4.

> Hint: You need to generalise to all possible examples to write a proof.

**3** Write identities for

**a** $3x + 6$        **b** $4x(x + 3) - 2(x + 3)$.

**4** Take any number.

**a** Multiply the two numbers on either side of your number.

**b** Add 1 to your answer.

**c** Take the square root of this answer.

**d** What do you notice?

**e** Prove this is true for any number.

**5 a** Find an identity for the product of any two numbers subtracted from the sum of their squares.

**b** Multiply your answer by the sum of the two numbers.

**c** Prove that this new answer is the same as the cubes of the two original numbers.

Q 1150, 1247   SEARCH

# .4 Expanding and factorising 2

**1** Expand and simplify these expressions.

**a** $(x + 3)(x + 4)$ **b** $(y + 4)(y - 1)$

**c** $(3 + p)^2$ **d** $(x + 5)(x - 5)$

**e** $(h - 5)(h + 2)$ **f** $(2x + 3)(x + 5)$

**g** $(3t + 1)(2t + 5)$ **h** $(3m + 4)^2$

**i** $(3p + q)(2p - 3q)$ **j** $(4x - 3y)^2$

**2 a** Show clearly that $(x - y)(x + y) \equiv x^2 - y^2$.

**b** Hence, without using a calculator, evaluate $5.1^2 - 4.9^2$.
You **must** show your workings.

**3** Copy and complete these identities.

**a** $\Box - 7x - x^2 \equiv (2 \Box x)(x + 9)$

**b** $15 + \Box - 2x^2 \equiv (\Box - x)(2x + 5)$

**c** $19x - 10 - \Box \equiv (5 - 2x)(3x - \Box)$

**4** Factorise these expressions.

**a** $x^2 + 8x + 7$ **b** $x^2 + 5x - 24$ **c** $x^2 - 5x + 4$

**d** $x^2 - 5x - 24$ **e** $y^2 - y - 12$ **f** $x^2 - 3x - 18$

**g** $x^2 + 4x - 12$ **h** $x^2 - 2x - 35$

**5** Factorise completely these expressions.

**a** $x^2 + 6x + 5$ **b** $x^2 + 7x - 18$ **c** $x^2 - 3x - 18$

**d** $x^2 - 15x - 100$ **e** $x^2 - 21x + 110$ **f** $x^2 - 16x + 64$

**g** $3x^2 + 21x + 30$ **h** $x^3 - 6x^2 - 40x$

**6** Factorise each of these expressions using double brackets.

**a** $2x^2 + 11x + 5$ **b** $3x^2 + 11x + 6$ **c** $5x^2 + 6x - 8$

**d** $6x^2 + 11x + 3$ **e** $12x^2 - 23x + 5$ **f** $15x^2 + 34x + 15$

**7** Factorise these expressions.

**a** $x^2 - 64$ **b** $m^2 - 25$ **c** $49 - t^2$

**d** $a^2 - b^2$ **e** $9y^2 - 100$ **f** $x^2 - \dfrac{1}{9}$

**g** $25a^2 - \dfrac{4}{9}$ **h** $2a^3 - 32a$

**1** Prove that the product of two odd numbers is always odd.

> Hint: Let the two numbers be $2n + 1$ and $2m + 1$.

**2** Given that two consecutive numbers can be written as $2n$ and $2n + 1$, prove that the sum of the squares of any two consecutive numbers is an odd number.

**3** Expand and simplify these expressions.

    **a** $(3x + 4)(x - 5)$     **b** $(3x - y)^2$.

**4**  **a** Expand and simplify $(p + q)^2$.

    **b** Hence, find the value of

       $2.84^2 + 2 \times 2.84 \times 1.16 + 1.16^2$.

**5**  **a** Factorise $a^2 - b^2$ into double brackets.

    **b** Use your answer to part **a** to work out $7.8^2 - 2.2^2$.

**6** Factorise these expressions.

    **a** $x^2 - 9$       **b** $p^2 - 49$       **c** $4n^2 - 81$

    **d** $x^2 - \dfrac{4}{25}$     **e** $3a^3 - 12a$

**7** Factorise these expressions into double brackets.

    **a** $x^2 + 7x + 12$     **b** $x^2 + 6x - 16$

    **c** $x^2 - 4x - 45$     **d** $x^2 - 11x + 28$

    **e** $x^2 - 8x + 16$     **f** $x^2 - 49$

    **g** $x^2 - 2x - 120$     **h** $x^2 - 196$

**8** $f(x) = 4x + 2$ and $g(x) = x^2 - 1$
Find

    **a** $f(0)$    **b** $g(0)$    **c** $f(2)$    **d** $g(2)$

    **e** $f(-1)$    **f** $g(-3)$    **g** $fg(4)$    **h** $gf(1)$

# Measuring lengths and angles

**1** Using a protractor measure the bearing of *B* from *A* in these diagrams.

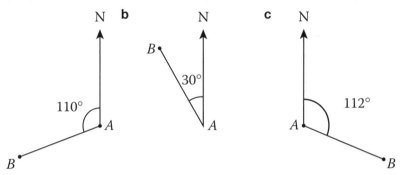

**a**

N

110°

*A*

*B*

**b**

N

*B*

30°

*A*

**c**

N

*A*

112°

*B*

**2** A ship sails on a bearing of 200° until it reaches its destination.
If it returns on the same route, what is the bearing of the
return journey?

**3** **a** Look at the diagram. Describe
the journey which begins at *A*
and finishes at *C* in terms of the
distance travelled and bearings.

**b** By scale drawing, find the
direct distance of *C* from *A*.

**4** The scale of a map is 1 : 5 000 000.

**a** How many km does 1 cm
represent?

**b** How many cm represent
14.5 km?

**c** How many km are
represented by 8.5 cm?

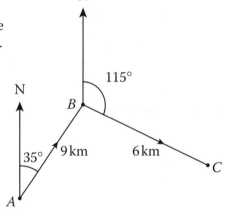

N

115°

N

*B*

35° 9 km

6 km

*C*

*A*

**5** Why would a scale 1 : 6 000 000 not be appropriate if Max wants
to do a scale drawing of his house? Suggest a more appropriate scale,
explaining your answer carefully.

**1**   Calculate the area of each shape.

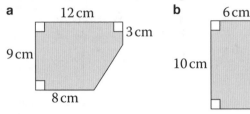

**a**

12 cm

3 cm

9 cm

8 cm

**b**

6 cm

10 cm

3 cm

4 cm

7 cm

15 cm

**2**   Calculate the area of each shape.

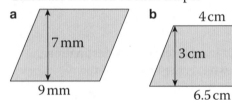

**a**

7 mm

9 mm

**b**

4 cm

3 cm

6.5 cm

**3**   The area of this rectangle is twice the area of the square.
Find the length of a side of the square.

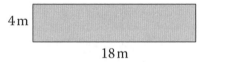

4 m

18 m

**4**   Calculate the area of this cross-section of Jacob's toy rocket.

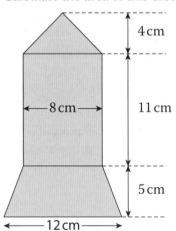

4 cm

8 cm

11 cm

5 cm

12 cm

# 7.3 Transformations 1

**1** Write vectors to describe the translation

    **a** triangle *A* to triangle *B*

    **b** triangle *A* to triangle *C*

    **c** triangle *B* to triangle *D*

    **d** triangle *C* to triangle *D*

    **e** triangle *D* to triangle *E*

    **f** triangle *E* to triangle *C*.

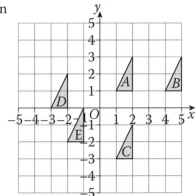

**2** Describe fully the transformation that maps

    **a** triangle *A* to triangle *B*

    **b** triangle *C* to triangle *B*

    **c** triangle *B* to triangle *D*

    **d** triangle *D* to triangle *A*

    **e** triangle *E* to triangle *A*.

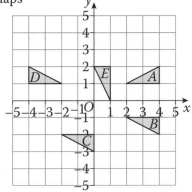

**3**  **a** Rotate flag A 90° clockwise about the origin. Label the image *B*.

    **b** Reflect flag *B* in the line $x = 0$. Label the image *C*.

    **c** (**Challenge**) Describe fully the transformation that maps flag *A* onto flag *C*.

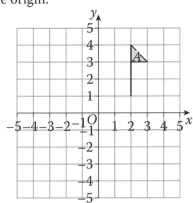

# 7.4 Transformations 2

**1** Copy this diagram.

**a** Enlarge rectangle *A* by scale factor 3, centre (0, 0). Label the image *B*.

**b** Write the perimeter of

   **i** rectangle *A*

   **ii** rectangle *B*.

**c** How many times larger is the perimeter of rectangle *B* compared to the perimeter of rectangle *A*?

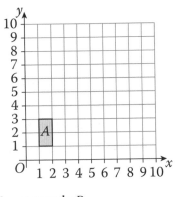

**d** Write the area of **i** rectangle *A* **ii** rectangle *B*.

**e** How many times larger than the area of rectangle *A* is the area of rectangle *B*? Write this value as a power of 3.

**2** Describe fully the single transformation that maps

**a** trapezium *X* onto trapezium *Y*

**b** trapezium *Y* onto trapezium *X*.

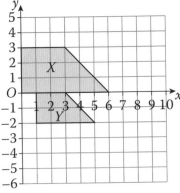

**3** Copy this diagram.

**a** Reflect *ABCD* in the *x*-axis, and label the image *A'B'C'D'*.

**b** Rotate *A'B'C'D'* by 180° about the origin, and label the image *A"B"C"D"*.

**c** Find the single transformation that maps *ABCD* to *A"B"C"D"*.

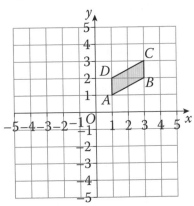

Q 1125   SEARCH

**1** The area of each shape is given. Calculate the unknown length.

**a**

? m

6 m

Area = 24 m²

**b**

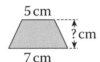

5 cm

? cm

Area = 40 cm²

**c**

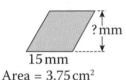

? mm

15 mm

Area = 3.75 cm²

**d**

5 cm

? cm

7 cm

Area = 48 cm²

**2** **a** Reflect shape L in the line x = 0.
Label the image M.
**b** Reflect shape M in the line x = 3.
Label the image N.
**c** Describe fully the single transformation that maps shape L onto shape N.

**3** Copy this diagram, and extend both axes up to 8.
Enlarge the shaded kite by scale factor 2, centre O.

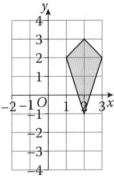

**4** Describe fully the single transformation that maps
**a** triangle A onto triangle B
**b** triangle B onto triangle A.

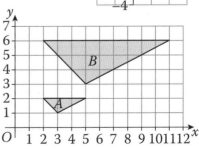

**1**  80 students are asked to choose their favourite season.

|  | Spring | Summer | Autumn | Winter | Total |
|---|---|---|---|---|---|
| Skiers | 8 | | | 12 | |
| Non-skiers | | 18 | 7 | 8 | |
| Total | 20 | | 16 | | 80 |

    **a**  Copy and complete the table.

    **b**  A student is chosen at random. Find the probability the student

       **i**  prefers summer  **ii**  is a skier who prefers winter.

**2**  You roll a biased
dice 150 times.
The table shows
the outcomes.

| Score | 1 | 2 | 3 | 4 | 5 | 6 |
|---|---|---|---|---|---|---|
| Frequency | 50 | 24 | 22 | 20 | 26 | 8 |

    **a**  If you roll the dice once more, estimate the probability that it will
land on a

       **i**  1     **ii**  6     **iii**  2 or 5.

    **b**  You plan to roll the dice 300 times. How many times would you
expect the dice to land on a

       **i**  2     **ii**  3 or 4?

**3**  Cameron carries out a survey about the words in a book. He chooses
a page at random and counts the letters in the first 150 words on that
page. The table shows the outcomes of his experiment.

| Number of letters | 1 | 2 | 3 | 4 | 5 | 6 | 7 | 8 |
|---|---|---|---|---|---|---|---|---|
| Frequency | 8 | 14 | 35 | 45 | 30 | 10 | 5 | 3 |

The book has 30 000 words.
Estimate the number of 3-letter words in the book.

**4**  A biased dice is in the shape of a
tetrahedron. You roll the dice
100 times. The table shows
the outcomes of this experiment.

| Score | 1 | 2 | 3 | 4 |
|---|---|---|---|---|
| Frequency | 48 | 19 | 16 | 17 |

    **a**  If you roll the dice once more, find the probability that it will land
on a  **i**  1  **ii**  3 or 4  **iii**  not 2.

    **b**  You plan to roll the dice 300 times. How many times would you
expect the dice to land on a  **i**  2  **ii**  1 or 4?

Q 1211, 1264  SEARCH

# Theoretical probability

**1** Suzanne decides to roll a biased dice 500 times.
The probability that the dice will land on a six is 0.45.
Work out an estimate for the number of times that the dice will land on a six.

**2** Freya carries out a statistical experiment. She throws a dice 120 times.
She scores a three 40 times.
Is the dice fair? Explain your answer.

**3** A dice is rolled 30 times in order to test its fairness.
The results of this experiment are shown below.

| 1 | 3 | 3 | 5 | 6 | 2 |
| 4 | 4 | 5 | 2 | 4 | 2 |
| 1 | 4 | 5 | 1 | 6 | 4 |
| 4 | 4 | 2 | 3 | 5 | 4 |
| 6 | 4 | 1 | 5 | 6 | 4 |

   **a** Work out the relative frequency of rolling each number.
   **b** Is the dice biased? Explain your answer.
   **c** What could you do to improve the experiment?

**4** Jordan can't decide where to get married.
She writes the names of 25 locations on individual cards,
numbered 1 to 25, and puts them in a bag.
   **a** What is the probability that a card, drawn at random, will be
      **i** a multiple of 5
      **ii** a prime number
      **iii** a factor of 24?
   **b** What is the probability that a card, drawn at random, will not
      **i** have a 1 in it
      **ii** be a square number?

## 8.3 Mutually exclusive events

**1** A bag contains red, blue and yellow balls.
The probability of selecting a red ball is 0.15.
The probability of selecting a blue ball is 0.4.
   **a** What is the probability of selecting a yellow ball?
   **b** What is the probability of selecting a ball that is not red?
   **c** If there are 20 balls in the bag, how many blue balls are there?

**2** The probability that Ethan walks to school is $\frac{5}{8}$ and the probability that he takes the bus is $\frac{1}{4}$. At all other times, Ethan's parents drive him to school in their car. Find the probability that
   **a** Ethan's parents drive him to school
   **b** Ethan does not walk to school
   **c** Ethan takes either the bus or walks to school.

**3** The probability of rain in Derby on Tuesday is 0.65.
What is the probability that Tuesday in Derby will be dry?

**4** A child's pushchair is available in five different fabrics. These fabrics and the probability with which they are selected by a consumer are shown in the table.

| Neutrals | Checks | City Chic | Vibrant | Pure |
|----------|--------|-----------|---------|------|
| 0.42     | 0.11   | $x$       | $2x$    | 0.05 |

Find the missing value $x$ and hence the missing probabilities.

**5** Georgia plays a game on the computer. The probability that she completes each level of the game is shown in the table.

| Level 1 | Level 2 | Level 3 | Level 4 |
|---------|---------|---------|---------|
| $8x$    | $4x$    | $2x$    | $x$     |

   **a** Comment on the difficulty of each level in comparison to the previous one.
   **b** Work out the probability of completing each level.

**6** A child has a set of building blocks, numbered 1 to 20, in a trolley.
The child selects one block at random.
   **a** Find the probability that the number on the chosen block is
      **i** an 8        **ii** a multiple of 4        **iii** a factor of 6
      **iv** a prime number     **v** the number 3 or greater than 15.
   **b** What can you say about the pair of events in part **v**?

1262, 1263 **SEARCH**

1   You throw a biased dice in the shape of a tetrahedron 100 times.
    The table shows the outcome of the experiment.

| Score | 1 | 2 | 3 | 4 |
|---|---|---|---|---|
| Frequency | 48 | 14 | 16 | 22 |

   a   You throw the dice again. Find the probability that it will land
       on a 2 or 3.

   b   If you throw the dice 500 times, how many times would you expect
       it to land on a 1?

2   Joan carries out a statistical experiment.
    She throws a dice 100 times. She scores a five 30 times.
    Is the dice fair? Explain your answer.

3   A credit card company offers a choice of designs for their cards.
    The table shows the picture on each card and the related
    probability.

| Shoes | Diamonds | Tiger | Sports car |
|---|---|---|---|
| 0.21 | $x$ | 0.28 | 0.19 |

   Find the value of $x$.

4   You put twenty balls, numbered 1 to 20, in a box. You select a ball at
    random.

   a   What is the probability that the number on the chosen ball is

       i    a 3                          ii    divisible by 5

       iii  a multiple of 4             iv    greater than 10

       v    a prime number             vi    not an odd number

       vii  a 4 or less                viii  a 7 or an even number?

   b   What can you say about the pair of events 'the number on the ball
       is a 7 or the number on the ball is even'?

# 9.1 Estimation and approximation

1 Write a suitable estimate for each of these calculations.

   a $\dfrac{21.3 \times 689}{73.6}$      b $\dfrac{231 \times 49}{(18.6)^2}$

2 By rounding all of the numbers to 1 significant figure, estimate the answer to each of these calculations.

   a $56 \times 923$      b $15\,048 \div 456$

   c $704 + 2520 \div 105$   d $\dfrac{52.3 - 12.8}{2.93 + 5.06}$

   e $\dfrac{32.71 \times 4.09}{2.15^2}$      f $\dfrac{3.93 - 2.57}{12.5 \times 2.6}$

   > Hint: Don't forget to use BIDMAS!

3 Write a suitable estimate for each of these calculations. In each case clearly show how you estimated your answer.

   a $\dfrac{39.9 \times 21.5}{1.98^3}$      b $\dfrac{47.9 \times 9.8^2}{0.49 \times 21.56}$      c $\sqrt{98.7 \div 1.98}$

   d $\{2.6^2 + (4.57 - 0.62)\}^2$           e $\dfrac{219 + (3.98 + 16.08)^2}{\sqrt{74.5 \div 2.11}}$

4 By rounding each of the values in these calculations to 1 significant figure, find an approximate answer.

   a $\dfrac{86.7 - 21.4}{5.95 + 4.18}$      b $\dfrac{42.71 \times 0.099}{2.03^3}$

   c $\sqrt{\dfrac{18.2^2}{0.82 \times 4.56}}$      d $\dfrac{3.1^4}{0.088 - 0.008\,76}$

5 Explain why rounding each of the values in these calculations to 1 significant figure would *not* be an appropriate method for estimating the results of the calculations.

   a $\dfrac{21.4 \times 5.15}{0.84 - 0.75}$      b $(4.56 - 4.32)^2$

Q 1005, 1043, 1057    SEARCH

# Calculator methods

**1** Calculate

**a** $4 \times 3^2 - 7 + 2$

**b** $18 - 7 \times 2^2 + 3^3$

**c** $\dfrac{5^2 + 3(10 - 7) + 2 \times 3}{5.72 - 3.22}$

**d** $\dfrac{4.9 \times 3.2}{6}$

**2** Calculate

$$\sqrt{0.5^2 + 1.4^2 - 2 \times 0.5 \times 1.4}$$

**3** Calculate these giving all the figures on your calculator display.

**a** $\dfrac{6.53^2 \times 2.19 + 7.34}{5.13 - 3.78}$

**b** $\dfrac{94.39 - (4.8 + 2.71)^2}{5.81^2 - 5.42}$

**4** Calculate these giving your answers to 4 significant figures.

**a** $\dfrac{3.241 \times 5.016}{2.897}$

**b** $\dfrac{9.194}{3.127 \times 1.563}$

**c** $\dfrac{1}{0.231} - \dfrac{8.254}{4.176}$

**d** $\sqrt{\left(\dfrac{2.564}{1.112}\right)}$

**e** $\dfrac{\sqrt{9.332} + (7.214)^3}{(8.246)^2}$

**5** Calculate these quantities.

**a** $\left(4\dfrac{4}{5}\right)^2$

**b** $2^8$

**c** $\sqrt{\dfrac{4}{9}}$

**d** $\sqrt[5]{243}$

**6** **a** Use approximations to estimate the value of $\dfrac{458 \times 0.036}{1.69}$.
Show all of your working.

**b** Use your calculator to find the exact answer.

**c** Calculate the percentage error for the approximation.

**7** Calculate these giving each answer to an appropriate degree of accuracy.

**a** $\sqrt{41.2^2 + 7.62^2}$

**b** $\sqrt{1.3^2 + 2.7^2 - 2 \times 1.3 \times 2.7}$

**c** $4\dfrac{1}{3} \div 1\dfrac{3}{7}$

**d** $\dfrac{-7 + \sqrt{7^2 - (4 \times 3 \times -6)}}{2 \times 3}$

# 9.3 Measures and accuracy

1 Convert these measurements to the units given.

a 30 mm = ___ cm          b 300 cm = ___ m

c 3 kg = ___ g            d 5000 ml = ___ litres

e 0.5 km = ___ m          f 6 litres = ___ ml

g 4.5 tonne = ___ kg      h 7 m = ___ cm

i 30 cl = ___ litres      j 0.25 cm = ___ mm

2 Write the upper and lower bounds of these measurements, which are all given to 3 significant figures.

a 9.48 m          b 38.2 kg          c 9.14 s

d 153 cm          e 19.5 g           f 60.7 m

3 Write the upper and lower bounds of each of these measurements, which have been measured to the degree of accuracy in brackets.

a 12 m (to the nearest m)          b 14.8 s (to 1 decimal place)

c 1200 g (to 2 significant figures)  d 24.5 kg (to 3 significant figures)

e 3.05 litres (to 2 decimal places) f 4.39 m (to the nearest cm)

4 Rupert decides to send a red rose to his girlfriend. The rose measures 30 cm, to the nearest centimetre. Rupert buys a presentation box of length 30.1 cm, measured to the nearest millimetre.

a Explain why the rose may not fit inside the box.

b What is the maximum length that Rupert may have to cut from the stem of the rose in order to make it fit in the box?

5 Find the upper and lower bounds of these calculations.

a The area of a rectangular pond with length 4 m and width 3 m, measured to the nearest centimetre.

b The range of temperature during one day in Derby if the maximum recorded temperature was 28 °C and the minimum recorded temperature was 18 °C, to the nearest degree Celsius.

c The density of a piece of metal if its mass is 1125 g, to the nearest gram, and its volume is 150 cm$^3$, to the nearest cubic centimetre.

Q 1006, 1067, 1121, 1246 **SEARCH**

**1** **a** Estimate the value of
$$\frac{89.4 \times 34.5}{1.92 \times 30.4}$$

**b** Calculate the value of
$$\frac{12.35 \times (3.4 + 4.9)}{2.4^2 \times 1.3}$$
Give your answer to 2 decimal places.

**2** Write the upper and lower bounds for these measurements to the degree of accuracy given.

**a** 4 m (nearest unit)

**b** 650 mm (nearest 10)

**c** 241.3 g (1 decimal place)

**d** 45 ml (nearest 5 ml)

**e** 8300 km (2 significant figures)

**f** 11.53 sec (2 decimal places)

**3** Write the upper and lower bounds of these calculations.

**a** The radius of a circle measured as 4.5 cm to 1 decimal place.

**b** The area of a rectangle with dimensions 12 cm and 6.8 cm taken to 2 significant figures.

**c** The width of a parallelogram which has an area of 20 cm² (to the nearest square centimetre) and a length of 8.45 cm (to 2 decimal places).

**d** The speed, in km/h, of a car which travels 35 km (to the nearest kilometre) in 30 minutes (to the nearest minute).

**4** Grace has a toothbrush holder that is 20.2 cm long, to the nearest millimetre. She buys a new toothbrush that measures 20 cm to the nearest centimetre. Explain why Grace's new toothbrush may not fit into her toothbrush holder.

# 10.1 Solving linear equations

**1** Solve these equations.

   **a** $3x + 7 = 25$                  **b** $2(4y - 1) = 18$

   **c** $5(6 - 4x) = 25$             **d** $3p + 5 = 32$

   **e** $19 = 17 - 2x$              **f** $22 = 4(5y - 2)$

   **g** $2 + 6x - 3 - 3x = 0$      **h** $4m - 9 = 23$

**2** Solve these equations with algebraic terms on both sides.

   **a** $7x + 5 = 3x - 7$           **b** $2x - 10 = 7x - 11$

   **c** $9 - p = 4p - 11$           **d** $3(y - 8) = 4(2y + 9)$

   **e** $2(3 - x) = 5(4x - 1)$      **f** $14m - 1 = 2m + 8$

   **g** $4(y + 8) - 3(y + 5) = 3(y + 1)$      **h** $6(x - 2) = 9x - 3(2x - 1)$

**3** Solve these equations.

   **a** $2x + 9 = 3x + 4$           **b** $7p - 5 = 5p + 1$

   **c** $4(2y + 1) = 6y + 8$         **d** $9a - 4 = 2(3a + 4)$

   **e** $10 - 4x = 6x - 5$          **f** $5y + 7 = 9 + 7y$

   **g** $4t - 3 = 12 - t$            **h** $4(3 - 4x) = 2(5 - 6x)$

**4** Solve these equations involving fractions.

   **a** $\dfrac{x + 9}{3} = \dfrac{2(x + 1)}{4}$        **b** $\dfrac{6p}{5} = \dfrac{8p - 4}{3}$

   **c** $\dfrac{y}{3} + \dfrac{2y}{5} = 11$          **d** $\dfrac{m}{3} + \dfrac{1}{4} = \dfrac{m}{4}$

**5** Solve these one-sided equations by algebraic methods.

   **a** $3x + 8 = 23$    **b** $5(y - 1) = 30$    **c** $\dfrac{p}{3} - 4 = 1$

   **d** $3t + 13 = 94$    **e** $4(3c + 5) = 8$    **f** $\dfrac{7x - 6}{10} = 5$

   **g** $\dfrac{4y - 4}{15} = 4$    **h** $\dfrac{10}{x} + 3 = 8$

**6** In 4 years' time Clare will be twice the age she was 10 years ago. Use an algebraic approach to determine Clare's age.

**7** If I add together one quarter of my age and one sixteenth of my age I get 10. If my age is $x$, form an equation in $x$ and solve to find the value of my age.

# 0.2 Quadratic equations

**1** Solve these quadratic equations by factorising.

   **a**   $x^2 + 6x + 8 = 0$      **b**   $x^2 + 11x + 24 = 0$

   **c**   $x^2 + 5x - 6 = 0$      **d**   $x^2 + 6x - 16 = 0$

   **e**   $x^2 = 9x - 20$         **f**   $x^2 + 7 = 8x$

   **g**   $x^2 - 4x = 0$          **h**   $2x^2 + 12x = 0$

> Hint: Parts **g** and **h** factorise into single brackets.

**2** Solve these quadratic equations by completing the square.

   **a**   $x^2 + 10x + 21 = 0$    **b**   $x^2 + 6x = 0$

   **c**   $x^2 - 8x + 1 = 0$      **d**   $x^2 + 5x - 3 = 0$

**3** Solve these quadratic equations by using the quadratic formula.
Give your answers to 3 significant figures.

   **a**   $x^2 + 5x + 1 = 0$      **b**   $2a^2 - 4a + 1 = 0$      **c**   $3x^2 + 7x - 2 = 0$

   **d**   $3p^2 = 6p + 2$         **e**   $3x - 7x^2 + 5 = 0$      **f**   $5t^2 + 3 = 9t$

**4** Solve these by writing them as quadratic equations.

   **a**   $x^2 + 4 = 5x$         **b**   $(x - 3)^2 = 10$

**5** Solve these equations, giving your answers to 3 significant figures.

   **a**   $x + \dfrac{2}{x} = 7$      **b**   $x - \dfrac{3}{x} = 5$      **c**   $\dfrac{2}{x + 3} + \dfrac{1}{2x + 1} = 4$

**6** The perpendicular height of a triangle is 3 cm less than
the length of its base. The area of the triangle is 27 cm².
Form a quadratic equation to represent the information
and solve the equation to find the dimensions of the triangle.

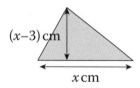

**7** Write $4x + 8 = \dfrac{25}{x}$ as a quadratic equation and solve it

   **a**   by using the quadratic formula     **b**   by completing the square.

# 10.3  Simultaneous equations

**1**  Solve these simultaneous equations by either adding or subtracting to eliminate one variable.

**a**  $x + 2y = 7$
  $x + y = 4$

**b**  $5x + 2y = 8$
  $2x + 2y = 2$

**c**  $p - 3q = 10$
  $4p + 3q = 10$

**d**  $a = 5b + 8$
  $3b = 8 - a$

**2**  Solve these simultaneous equations by the elimination method.

**a**  $3x + y = 11$
  $2x + 2y = 10$

**b**  $4x - y = 9$
  $3x + 2y = 4$

**c**  $5a + 3b = 19$
  $3a - 2b = 19$

**3**  Solve these simultaneous equations by the elimination method.

**a**  $2x + 3y = 21$
  $x - y = 8$

**b**  $2p + 5q = 1$
  $3p - 2q = 11$

**c**  $3s - 7t = 27$
  $5s + 3t = 1$

**d**  $y = 4 - x$
  $3x + 44 = 5y$

**e**  $2y = 18 - 4x$
  $0 = 3x - 2y + 4$

**f**  $4a - 2b + 2 = 0$
  $3b - 3a + 3 = 0$

**4**  Use the diagram to solve these pairs of simultaneous equations.

**a**  $x + y = 7$
  $y = 2x + 1$

**b**  $y = 2x - 3$
  $y = 2 - 3x$

**c**  $y = 2x + 1$
  $y = 2 - 3x$

**d**  Using the diagram, explain why $y = 2x + 1$ and $y = 2x - 3$ have no solutions.

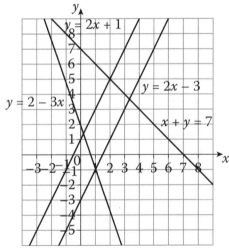

**5**  Solve these simultaneous equations.

**a**  $x^2 + y = 88$
  $y = 7$

**b**  $6y - x^2 = 23$
  $y = 8$

**c**  $y = x^2 + 3$
  $y = 3x + 7$

**d**  $p = 3q^2$
  $p = 7q - 2$

Q 1174, 1177, 1236, 1319  **SEARCH**

# 0.4 Approximate solutions

**1** **a** Show that the equation $x^3 - x^2 = 150$ can be rearranged to $x = \sqrt[3]{150 + x^2}$.

**b** This process can be used to find an approximate solution to $x^3 - x^2 = 150$.

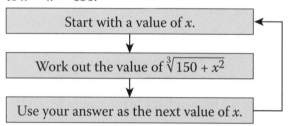

Use the process four times. Start with $x = 5$.
Give your answer to 3 dp.

**c** Substitute your answer to part **c** into $x^3 - x^2$.
Comment on the accuracy of your solution to the equation $x^3 - x^2 = 150$.

**2** **a** Show that the equation $m(m + 1) = 50$ can be rearranged to

   **i** $m = \sqrt{50 - m}$          **ii** $m = 50 - m^2$

**b** Use this process to find an approximate solution to $m(m + 1) = 50$.

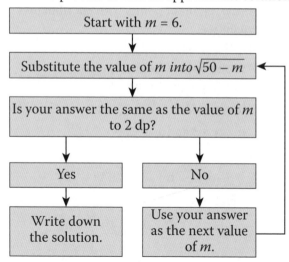

**c** Try to repeat the process by substituting into $50 - m^2$.
What do you notice?
Is your value of $m$ a solution to the equation?

# 10.5 Inequalities

1. Solve these linear inequalities, representing your solution on a number line.

    a $2x < 9$

    b $3x + 5 \leq 23$

    c $5x - 4 \geq 11$

    d $-4x > 16$

    e $3(x - 8) < 6$

    f $2x + 4 \geq 5(x - 1)$

    g $10 \leq \dfrac{x}{3}$

    h $3(x - 1) > 7(x + 3)$

2. $n$ is an integer.

    a List all the values of $n$ which satisfy the inequality

    $-1 \leq n < 4$.

    b Solve the inequality

    $4x + 3 \geq 5$.

3. Solve these inequalities.

    a $2x - 5 > 17$

    b $3(x + 1) \leq 2x$

    c $1 + n \geq -3$

    d $3 - 2x \leq 9$

    e $\dfrac{x}{5} \geq 7$

    f $\dfrac{2x}{3} - 1 < 5$

    g $24 < 8x < 3x + 20$

    h $x^2 \leq 64$

4. Draw a diagram to show the region of points with coordinates that satisfy the four inequalities

    $y \geq 0, x \geq 0, y \leq 5 - x$ and $y \geq x - 1$.

5. Draw suitable diagrams to show these inequalities. Remember to leave the required region *unshaded*.

    a $x \leq 5$

    b $y \geq -1$

    c $x < 5$ and $y < 4$

    d $-3 \leq x \leq 2$

    e $x > 3$ and $x < -2$

6. Solve these inequalities.

    a $3(p - 2) > 5p$

    b $7p \leq 2(4p + 7)$

7. Solve these quadratic inequalities.

    a $x^2 > 81$

    b $x^2 + 7x + 18 < 0$

**1** Solve these double-sided equations.

   **a**  $4(x + 3) = 5(x + 2)$         **b**  $8(2y + 1) = 3(7y - 4)$

   **c**  $5(1 - p) = 3(2p + 9)$       **d**  $3(5 - 2q) = 7(1 - 2q)$

   **e**  $5a + 2(7a - 3) = 51$        **f**  $4x - 3(10 - 2x) = 20$

   **g**  $(x + 2)(x + 5) = (x + 3)^2$    **h**  $(y - 3)^2 = (y + 4)^2$

**2** Solve these equations involving fractions.

   **a**  $\dfrac{x + 4}{2} = \dfrac{3(x + 6)}{5}$        **b**  $\dfrac{2x}{5} = \dfrac{4(x - 1)}{9}$

**3** Solve these quadratic equations by factorising.

   **a**  $x^2 + x - 12 = 0$          **b**  $x^2 + 17x + 30 = 0$

   **c**  $3x^2 - 23x - 8 = 0$        **d**  $4x^2 - 1 = 0$

**4** Solve these quadratic equations using the quadratic formula, giving your answers to 3 significant figures.

   **a**  $x^2 + 6x + 3 = 0$          **b**  $5z^2 - 10z - 3 = 0$

   **c**  $2a^2 = 7a - 2$             **d**  $3s^2 + 5 = 9s$

**5** Solve these simultaneous equations algebraically.

   **a**  $x + y = 6$     **b**  $2x + y = 3$     **c**  $2x - 3y = 2$

       $2x - y = 9$       $3x + 3y = 3$       $5x + 2y = 24$

**6** The length of this rectangle is 5 cm more than its width.
The area of this rectangle is 33.44 cm².
What are its length and width?

Write an equation to solve this problem.
Use interval bisection to find the width of
the rectangle to 3 decimal places.

$x$

$x + 5$

**7** Solve these inequalities.

   **a**  $5x - 9 < 11$     **b**  $3x + 1 \geq 5x - 3$    **c**  $y^2 \leq 9$

   **d**  $2(3 - y) \geq 3(y + 3)$   **e**  $2x^2 - 3 < 29$

In this exercise, give your answers to 3 significant figures where appropriate.

**1**   Find the circumference of these circles.

**a**
4 cm

**b**
1.9 mm

**c**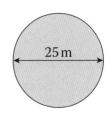
25 m

**2**   Find the area of the circles in question **1**.

**3**   A circle has an area of 98 mm².

     Find the circumference of the circle.

**4**   Jason has a lawn which is a
     rectangle with a semi-circle
     at each end.

     Find the area and perimeter
     of Jason's lawn.

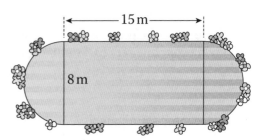
15 m

8 m

**5**   A circular flowerbed of radius 1.4 m is
     surrounded by a path of width 70 cm.

     Find the area of the path.

**6**   Ashim draws this picture on a piece of
     paper measuring 12 cm by 8 cm.
     He colours the six circles red and the rest
     of the paper blue.

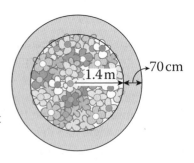
1.4 m
70 cm

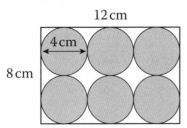
12 cm
4 cm
8 cm

Calculate the area of the paper that is blue.

**1.2** **Circles 2**

**1** Find the area of each sector.

**a**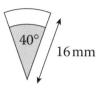
40° 16 mm

**b**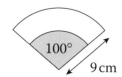
100° 9 cm

**c**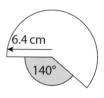
6.4 cm
140°

**d**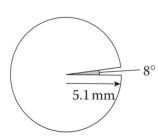
8°
5.1 mm

**2** Find the perimeter of each of the sectors in question **1**.

**3** Find each of the shaded angles.

**a** arc length = 4.2 cm

5 cm

**b** arc length = 24 m
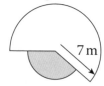
7 m

**c** Area = 65 m²
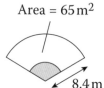
8.4 m

**d** arc length = 58.9 mm
75 mm

**4** A wall mirror is shaped as follows:
ABCD is a rectangle where

AB = DC = 60 cm and
AD = BC = 85 cm.

AOB is a sector of a circle of radius 0.425 m.
Angle AOB is 90°.
Calculate the perimeter of the mirror.

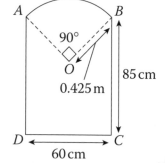
A                    B
90°
O
0.425 m            85 cm
D                    C
60 cm

# 11.3 Circle theorems

1 Find the missing angles, giving reasons for your answers.

a
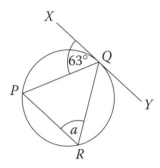

b
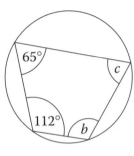

2 Find the angles marked with letters.

a
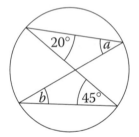

b
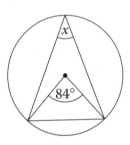

3 Find the missing angles, giving reasons for your answers.

a
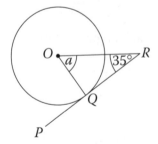

b
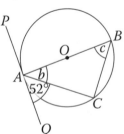

4 Prove that triangles *OBA* and *OCA* are congruent.

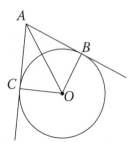

Q 1087, 1142, 1321   SEARCH

# 1.4 Constructions and loci

**1** **a** Draw a line 6 cm long and label it *AB*. Construct the perpendicular bisector of *AB* and label it *CD*. Label the point where these two lines meet as *X*.

   **b** Bisect the angle *CXB*. Label this new line *XY*.

   **c** What is the size of angle *CXY*?

**2** **a** Construct an angle of 60°.

   **b** Bisect this angle to give an angle of 30°.

**3** **a** Trace this line and points marked *x* and *y*.

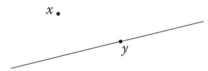

   **b** Use a ruler and a pair of compasses to construct a perpendicular from the point *x to* the line.

   **c** Use a ruler and a pair of compasses to construct a perpendicular from the point *y on* the line.

**4** **a** Copy the diagram of a garden *ABCD* using a scale of 1 cm to 1 m.

   **b** A sprinkler is placed at point *A*. The sprinkler waters the garden up to 5 m away as it rotates.

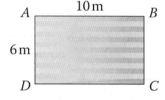

   A second sprinkler is placed at point *C*. It has a maximum reach of 4 m. Shade the area of the garden that remains unwatered.

**5** **a** Construct an isosceles triangle with sides 6 cm, 6 cm and 4 cm.

   **b** A garden is in the shape of an isosceles triangle as shown.

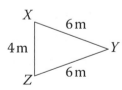

A water feature is to be placed in the garden so that it is

- nearer to *YZ* than *XY*

- closer to point *Y* than to point *Z*.

Use your diagram from part **a** to represent the area where the water feature may be placed. Shade this area.

**1** Calculate the **i** circumference **ii** area, of each circle.

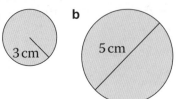

**a** 3 cm   **b** 5 cm

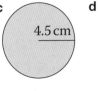

**c** 4.5 cm

**d** 3.2 cm

**2** **a** Find the area of this sector.   **b** Calculate the missing angle.

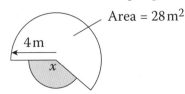

75° 3.6 cm

Area = 28 m²

4 m

x

**3** **a** Draw a line $AB$, so that $AB = 12$ cm.

$A$ ⎯⎯⎯⎯⎯⎯⎯⎯ $P$ ⎯⎯⎯⎯ $B$

**b** Mark the point $P$, so that $AP = 8$ cm.

**c** Construct the perpendicular to $AB$ that passes through point $P$.

**4** **a** $A$ and $B$ are two points 8 cm apart. Draw the locus of points that are equidistant from $A$ and $B$.

**b** On the same diagram, draw the locus of points that are 3 cm from $A$.

**5** $A$, $B$ and $C$ are points on the circumference of a circle, centre $O$.

**a** Find angle $AOC$.

**b** Give a reason for your answer.

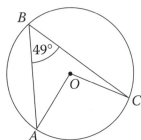

$B$ 49° $O$ $C$ $A$

**6** Find the missing angles.

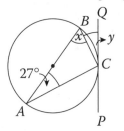

$B$ $Q$
$x$ $y$
$C$
27°
$A$ $P$

# Proportion

**1** This table gives information concerning the number of students in Years 7 to 9 that stay for a school dinner on 7th October.

|  | Year 7 | Year 8 | Year 9 |
|---|---|---|---|
| **Boys** | 77 | 65 | 58 |
| **Girls** | 63 | 60 | 57 |

    **a** How many students in Years 7 to 9 had a school dinner on 7th October?

    **b** Work out the percentage of students that were

        **i** girls     **ii** Year 7 boys.

    **c** What percentage of the girls were in Year 8?

**2** Samantha spends $\frac{3}{5}$ of her monthly salary on tax, her mortgage and her regular bills. She spends $\frac{1}{6}$ of her monthly salary on food bills and $\frac{1}{12}$ on her car payments. The rest she spends on shoes.

    **a** What proportion of her monthly salary does Samantha spend on shoes?

    **b** If Samantha earns £2400 a month, write down the amounts that she pays for

        **i** tax, mortgage and bills     **ii** shoes.

**3** Kent surveys his team of 30 engineers. He notes that 25 of his team really enjoy their job and that 4 of his team have worked in other professions before becoming engineers.

    **a** What proportion of the engineers really enjoy their job?

    **b** What proportion of Kent's team have only ever worked as engineers?

# 12.2 Ratio

**1** Simplify these ratios.

  **a** £8 : £4           **b** 50 kg : 5 kg

  **c** 2 m : 20 000 m     **d** 54p : 36p

  **e** 21 cm : 35 cm : 14 cm   **f** £3 : 60p

  **g** 1.25 kg : 750 g      **h** 625 m : 1.4 km

  **i** 25 min : 1.25 hours   **j** 1.2 m : 45 cm : 2.25 m

**2** Simplify these ratios.

  **a** $\dfrac{1}{2} : 2$   **b** $3 : \dfrac{1}{4}$   **c** $\dfrac{1}{2} : \dfrac{1}{5}$   **d** $\dfrac{2}{3} : \dfrac{4}{7}$

**3** Write these ratios in the form 1 : $n$.

  **a** 3 : 5   **b** 15 : 4   **c** 2 cm : 10 km  **d** 5 cm : 2 km

**4** Divide £480 in these ratios.

  **a** 2 : 1   **b** 1 : 5   **c** 5 : 3   **d** 7 : 5

**5** **a** Simplify these ratios.

    **i** 3 m : 3000 m   **ii** 25p : £3   **iii** 30 : 12 : 18   **iv** $\dfrac{1}{4} : \dfrac{2}{3}$

  **b** Find the missing values in these equivalent ratios.

    **i** $a : 3 = 8 : 12$   **ii** $b : 5 = 1 : 3$   **iii** $7 : c = 8 : 5$

**6** In a class, the ratio of girls to boys is 5 : 4. There are 15 girls in this class. Work out the number of boys.

**7** **a** Divide £350 in the ratio 3 : 4.   **b** Divide £950 in the ratio 8 : 11.

**8** Angela, Rajat and Ayesha have savings in the ratio 5 : 4 : 1. If Rajat has £2400, calculate the amount of money that Angela and Ayesha have in savings.

**9** In the UK in 2011 there were approximately 96 males for every 100 females in the population.

  **a** Write this as a ratio of males to females. Give your answer in its simplest terms.

  **b** What proportion of the population was female? Give your answer as a fraction in its simplest form.

  **c** If there were 63.2 million people in the UK in 2011, how many males were there? Give your answer to 3 significant figures.

Q 1036, 1038, 1039, 1103  **SEARCH**

1   Increase 432 m by 12%.

2   The gym membership fee increases by 12%. What is the new fee for

   a   standard membership, previously £20 per month

   b   gold membership, previously £32 per month

   c   platinum membership, previously £37.50 per month?

3   Simon is shopping for a new notebook computer. He decides on one
    particular brand and model and discovers that it is on offer in three
    different shops.

Find the cost of the notebook computer in all three shops and decide on
the best offer for Simon.

4   Julia bought her house for £250 000 in 2009. In 2014, Julia decided to
    sell her house and placed it on the market at £575 000. What was the
    percentage increase of the house price in the five years that Julia owned it?

5   In the summer sale, all clothes have a 20% discount.

   a   What is the sale price of a pair of trousers that cost £85 before
       the sale?

   In the last week of the sale, these sale prices are reduced by a further 20%.

   b   What is the price of the trousers now?

   c   What is the overall percentage reduction of the pair of trousers
       after both discounts have been applied?

6   Gemma invests £100 in a bank account.
    Simple interest of 4% is added at the end of each year.
    Work out how much money Gemma has at the end of

   a   the first year      b   three years.

7   Zac bought a van for £15 000.
    Each year the van depreciated by 5%.
    Work out the value of Zac's van 2 years after he purchased it.

**1** In a primary school class, $\frac{1}{5}$ of the students have black hair, $\frac{3}{10}$ of the students have blonde hair and $\frac{7}{15}$ of the students have brown hair. The rest have red hair.

  **a** What proportion of the class have red hair?

  **b** If there are 30 students in the class, how many have

    **i** black hair        **ii** brown hair?

**2** Write each of these ratios in the form 1: $n$.

  **a** 3 : 12        **b** 5 : 35        **c** 16 : 480

  **d** 65p : £ 4.55        **e** 35 g : 35 kg        **f** 0.18 : 5.76

**3** Split 360 into these ratios

  **a** 1 : 2        **b** 2 : 3        **c** 4 : 5

**4** In a class, the ratio of girls to boys is 4 : 3. There are 16 girls in this class. Work out the number of boys.

**5** Use a calculator to find each of these percentage changes.

  **a** Increase £36 by 22%        **b** Increase 15 m by 24%

  **c** Decrease $42 by 54%        **d** Increase 65 kg by 12%

  **e** Decrease 256 litres by 8%        **f** Decrease 952 km by 35%

**6** In a sale, a handbag costing £40 is reduced by 20%. What is the sale price?

**7** In a sale, all prices are reduced by 15%. A TV in the sale costs £679.15. What was the original price of the TV?

**8** Kevin and Kathleen decide to play the National Lottery. Kevin contributes £10.50 and Kathleen contributes £7.00 to buy tickets.

  **a** Write the ratio of Kevin's contribution to Kathleen's contribution in its simplest form.

Kevin and Kathleen win £1500.

  **b** Divide their prize money between them in the ratio of their original contributions.

# Factors and multiples

**1** Use a factor tree to write each of these numbers as a product of its prime factors.

    **a** 60     **b** 140     **c** 378     **d** 504     **e** 2156

**2** Write each of these numbers as a product of its prime factors.

    **a** 210           **b** 540

    **c** 1350         **d** 1750

    **e** 1694         **f** 4732

**3** Find the HCF and LCM of each pair of numbers.

    **a** 60 and 72     **b** 24 and 40     **c** 24 and 29

**4** Find the HCF and LCM of each set of three numbers.

    **a** 24, 36, 60     **b** 48, 72, 240     **c** 21, 63, 504

**5** **a** Express these numbers as products of their prime factors.

        **i** 750         **ii** 1470

    **b** Find

        **i** the HCF     **ii** the LCM  of 750 and 1470.

    **c** Find the smallest number that can be multiplied by 750 to give a square number.

**6** The area of a rectangular lawn is 90 m². By grouping the prime factors of 90, find all the possible dimensions of the lawn.

> Hint: Remember to include the dimensions where one length is 1 m.

**7** **a** At a disco it was discovered that Connor, Sunita, Nathan and Pierre had birthdays on the 6th, 12th, 15th and 24th of the month. Amanda joined the group and it was discovered that her birthday was the highest common factor of everyone else's. What day of the month was Amanda born on?

    **b** Later, Amanda's dad came to take her home and it was found that the number on his car registration plate was the lowest common multiple of the ages of Amanda and her four friends. What was the number on the registration plate?

## 13.2 Powers and roots

**1** Use your calculator, where necessary, to work out these powers and roots.

| | | | | | |
|---|---|---|---|---|---|
| **a** | $6^2$ | **b** | $14^2$ | **c** | $4^3$ |
| **d** | $1^3$ | **e** | $15^2$ | **f** | $(-6)^2$ |
| **g** | $7^3$ | **h** | $12^3$ | **i** | $10^3$ |
| **j** | $\sqrt{81}$ | **k** | $\sqrt{169}$ | **l** | $\sqrt[3]{64}$ |
| **m** | $\sqrt[3]{17\,576}$ | **n** | $\sqrt{4900}$ | **o** | $\sqrt[3]{1000}$ |

**2** Use the trial and improvement method to find the square root of each of these numbers to 1 decimal place. Record your results in a table. The first one has been started for you.

**a** $\sqrt{90}$

| Estimate | Check (square of estimate) | Answer | Too big or too small |
|---|---|---|---|
| 9 | $9^2$ | 81 | Too small |
| 10 | $10^2$ | 100 | Too big |
| 9.5 | | | |
| | | | |

**b** $\sqrt{70}$  **c** $\sqrt{120}$  **d** $\sqrt{250}$

**3** Find the value of each of these expressions.

**a** $(3^2)^3$  **b** $\left(\sqrt{4}\right)^2$  **c** $a^2 \times a^4$

**d** $\dfrac{3^4 \times 3^5}{3}$  **e** $5^0$

**4** Write

**a** 8 as a power of 2  **b** 64 as a power of 4

**c** 81 as a power of 3  **d** 100 000 as a power of 10

**e** 1296 as a power of 6  **f** $\dfrac{1}{4}$ as a power of $\dfrac{1}{2}$.

**5** Simplify these expressions.

| | | | | | |
|---|---|---|---|---|---|
| **a** | $2^5 \times 2^3$ | **b** | $4^3 \times 4^6$ | **c** | $8^6 \times 8$ |
| **d** | $x^2 \times x^7$ | **e** | $5^9 \div 5^4$ | **f** | $10^7 \div 10^6$ |
| **g** | $a^8 \div a^8$ | **h** | $9^4 \div 9^7$ | **i** | $3^2 \times 3^2 \times 3^2$ |
| **j** | $y^2 \times y^3 \times y^4$ | **k** | $6^5 \div 6^2 \times 6^4$ | **l** | $7^6 \times 7^4 \div 7^3$ |

Q 1033, 1053, 1924  **SEARCH**

**1**   Which of these numbers are irrational? Copy and complete the table.

| Number | ✓ or ✗ |
|--------|--------|
| $\sqrt{9}$ | |
| $5\pi$ | |
| $\frac{4}{7}$ | |
| $\sqrt[3]{5}$ | |
| $\sqrt{15}$ | |

**2**   Evaluate these expressions without using a calculator.

**a**   $\sqrt{3} \times \sqrt{3}$    **b**   $4 + \sqrt{4} \times \sqrt{4}$   **c**   $\sqrt{8} \times \sqrt{2}$    **d**   $\sqrt{5} \times \sqrt{20}$

> Hint: For parts **c** and **d** use the fact that
> $\sqrt{a} \times \sqrt{b} = \sqrt{a \times b}$.

**3**   Given that $\sqrt{3} \approx 1.73$ and $\sqrt{5} \approx 2.24$, find approximate values for these surds, without using a calculator. Show full working.

**a**   $\sqrt{12}$    **b**   $\sqrt{20}$    **c**   $\sqrt{45}$    **d**   $\sqrt{15}$

> Hint: Simplify each surd using the fact that
> $\sqrt{a \times b} = \sqrt{a} \times \sqrt{b}$.

**4**   Simplify these expressions.

**a**   $\sqrt{12}$    **b**   $\sqrt{18}$    **c**   $\sqrt{45}$    **d**   $\sqrt{63}$

**e**   $\sqrt{20}$    **f**   $\sqrt{50}$    **g**   $\sqrt{27} + 4\sqrt{3}$   **h**   $2\sqrt{24} - \sqrt{54}$

> Hint: Use the fact that $\sqrt{ab} = \sqrt{a \times b} = \sqrt{a} \times \sqrt{b}$.

**5**   Simplify these fractions.

**a**   $\dfrac{1}{\sqrt{3}}$    **b**   $\dfrac{1}{\sqrt{5}}$    **c**   $\dfrac{3}{\sqrt{7}}$

**d**   $\dfrac{1}{2\sqrt{3}}$    **e**   $\dfrac{\sqrt{7}}{3 - \sqrt{7}}$    **f**   $\dfrac{4 + \sqrt{3}}{2 - \sqrt{3}}$

1   Find the

   i    HCF      ii    LCM  of these pairs of numbers.

   a    36 and 60

   b    45 and 120

   c    54 and 504

2   1800 can be expressed by the products of prime factors $2^x \times 3^y \times 5^z$.

   Find the values of $x$, $y$ and $z$.

3   Work out

   a   $6^2$        b   $10^4$        c   $(-1)^5$

   d   $\sqrt{121}$     e   $\sqrt[3]{1000}$    f   $\sqrt[3]{27}$

4   Use the trial and improvement method to find the cube root of each of these numbers to 2 decimal places.
Record your results in a table.

   a   $\sqrt[3]{10}$     b   $\sqrt[3]{60}$     c   $\sqrt[3]{135}$     d   $\sqrt[3]{400}$

5   Say whether each of these numbers is rational or irrational.

   a   $\sqrt{49}$     b   $7\pi$     c   $\sqrt[3]{11}$

6   Evaluate these expressions without using a calculator.

   a   $\sqrt{2} \times \sqrt{8}$    b   $\sqrt{18} \times \sqrt{2}$    c   $\sqrt{5} \times \sqrt{20}$    d   $\sqrt{45} \times \sqrt{5}$

**4.1**    **Equation of a straight line**

**1**   Draw the graph of $y = 3x - 10$.

> Hint: Choose three values of $x$ and construct a table.
> Find the corresponding values of $y$. Think carefully about the axes.

**2**   For each of these straight-line graphs

   **i**    write the coordinate of the $y$-intercept

   **ii**   write the gradient of the line.

   **a**   $y = 3x + 4$       **b**   $y = 5x - 1$

   **c**   $y = 6 - 3x$      **d**   $x + y = 2$       Hint: $y = ?$

   **e**   $2y = 8 - x$      **f**   $4x - 2y = 3$

**3**   Write the equation of a line that is parallel to

   **a**   $y = 2x + 5$      **b**   $y = 4 - 3x$      **c**   $x + 2y = 1$.

**4**   Find the equation of a line that is parallel to $y = 4 - x$
and cuts the $y$-axis at $(0, 1)$.

**5**   Find the equation of a line that is parallel to $y = 3x + 6$
and passes through

   **a**   $(0, -2)$       **b**   $(2, 5)$       **c**   $(4, 1)$.

**6**   Write the equation of the line that passes through

   **a**   $(0, -2)$ and $(4, 0)$

   **b**   $(-1, 4)$ and $(2, -2)$

   **c**   $(3, 4)$ and $(-1, -2)$.

**7**   Find the equation of line $A$.

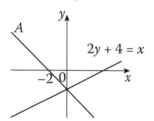

**8**   Sketch each of these line graphs on a separate set of axes.

   **a**   $y = 3x + 2$      **b**   $2y = x + 4$      **c**   $y = 5 - x$

# 14.2 Linear and quadratic functions

**1** A line has gradient 3 and $y$-intercept 4.

   **a** Plot the graph of this function.

   **b** Write the equation of the line.

**2** **a** Copy and complete this table to generate coordinates for the graph of $y = x^2 + 2x$.

| x | −3 | −2 | −1 | 0 | 1 | 2 | 3 |
|---|---|---|---|---|---|---|---|
| $x^2$ | | 4 | | | | | |
| 2x | | −4 | | | | | |
| y | | 0 | | | | | |

   **b** By drawing appropriate axes, plot the coordinates that you have found in part **a** and join them to form a smooth parabola.

**3** **a** Copy and complete this table to generate coordinates for the graph of $y = x^2 - x - 4$.

| x | −2 | −1 | 0 | 1 | 2 | 3 | 4 |
|---|---|---|---|---|---|---|---|
| $x^2$ | 4 | | | | | | |
| −x | 2 | | | | | | |
| −4 | −4 | | | | | | |
| y | 2 | | | | | | |

   **b** By drawing appropriate axes, plot the coordinates that you have found in part **a** and join them to form a smooth parabola.

**4** The point (2, 9) lies on which of these quadratic graphs?

   **A** $y = x^2 - 2x + 1$    **B** $y = x^2 - 3x - 10$    **C** $y = 2x^2 + 1$

   **D** $y = 3x^2 - 2x - 1$    **E** $y = x^2 + 4x - 3$

**5** **a** Copy and complete this table to generate coordinates for the graph of $y = x^2 - 2x - 3$.

| x | −2 | −1 | 0 | 1 | 2 | 3 | 4 |
|---|---|---|---|---|---|---|---|
| $x^2$ | | 1 | | | | | |
| −2x | | 2 | | | | | |
| −3 | | −3 | | | | | |
| y | | 0 | | | | | |

   **b** Solve these equations graphically.

   **i** $x^2 - 2x - 3 = 0$    **ii** $x^2 - 2x - 3 = -3$    **iii** $x^2 - 2x - 3 = x + 1$

Q 1180, 1312    SEARCH

**1** Plot the curve of $y = x^2 - x - 2$ for $-2 \le x \le 3$ and find the coordinates of its minimum point.

> Hint: Read off the $x$-coordinate for the minimum point. Substitute this value into $y = x^2 - x - 2$ to check the value of the $y$-coordinate.

**2** **a** Write $f(x) = 2x^2 - 16x + 9$ in the form $f(x) = a(x + b)^2 + c$ where $a$, $b$ and $c$ are to be determined.
    **b** Hence, give the minimum value of $f(x)$.
    **c** For what value of $x$ does this occur?

**3** Complete the square on these functions and hence write the minimum value of each function.
    **a** $f(x) = x^2 + 4x + 1$         **b** $f(x) = x^2 + 10x + 18$
    **c** $f(x) = x^2 - 4x - 5$         **d** $f(x) = 2x^2 - 4x - 3$

**4** For each of these quadratic graphs, find
    **i** the coordinates of the point where the graph cuts the $y$-axis
    **ii** the coordinates of the points where the graph cuts the $x$-axis
    **iii** the coordinates of the minimum (or maximum) point of each graph.
    **a** $y = x^2 + 4x - 5$         **b** $y = x^2 - 6x + 9$
    **c** $y = 2x^2 - x - 3$         **d** $y = 3 - 2x - x^2$

**5** Sketch each of the graphs in question **4**.

> Hint: Be careful with part **d**. Notice that the $x^2$ term is negative. What does this tell you about the graph?

**6** Sketch these quadratic graphs.
    **a** $y = x^2 + 2x - 8$         **b** $y = x^2 - 2x - 3$

> Hint: Find the coordinates of the points where each graph cuts the $x$- and $y$-axes. Complete the square and find the minimum point of each graph.

**7** The diagram shows a rectangle $x$ metres by $5 - x$ metres.
    **a** Write an expression for the area of the rectangle.
    **b** Let the area of the rectangle be $y\,m^2$.
        Plot a graph of $y$ against $x$.
    **c** Use your graph to find
        **i** the maximum possible area of the rectangle
        **ii** the dimensions of the rectangle if the area is $4\,m^2$.

$x$

$5 - x$

# 14.4 Kinematic graphs

1. Here is part of a distance–time graph of David's journey from his house to the shops and back.

   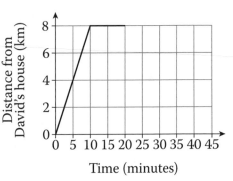

   **a** Work out David's speed for the first 10 minutes of the journey.

   David spent 10 minutes at the shops. He then travelled back to his house at 32 km/h.

   **b** Copy and complete the distance–time graph.

2. Finlay sets out at 1 p.m. from Farndon and walks at a speed of 6 km/h to Malpas, 12 km away.

   At 1.30 p.m., Joe jumps on his bike in Malpas and cycles to Farndon, reaching his destination in half an hour.

   **a** Construct a distance–time graph to show both journeys.

   **b** Work out the average speed at which Joe cycles.

   **c** Use your graph to work out when the two boys meet.

3. Aimée leaves her home in Clutton and cycles to Chester, 20 km away. On the outward journey, Aimée cycles at an average speed of 20 km/h but stops halfway for a 15 minute break. On reaching Chester, she spends 45 minutes at a friend's house trying to fix a flat tyre on her bicycle. She leaves her bike there in Chester and catches a bus back home. She arrives home in Clutton 25 minutes later.

   **a** Construct a distance–time graph to represent Aimée's journey.

   **b** How far did Aimée cycle before her break on the outward journey?

   **c** Work out the average speed of the bus in km/h.

**1** Write the equations of these three lines.

**a** Line A has a gradient of 3 and crosses the $y$-axis at $(0, 2)$.

**b** Line B has a gradient of 4 and passes through $(1, -1)$.

**c** Line C passes through $(0, 7)$ and $(3, -5)$.

**2** Write the equations of the three graphs labelled **a** to **c**.

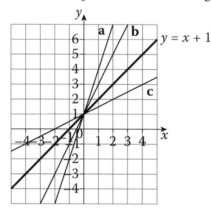

**3** Choose either 'parallel' or 'perpendicular' to complete these sentences.

**a** The line $y = 2x + 1$ is _____ to the line $y = 2x - 3$.

**b** The line $y = 6 - x$ is _____ to the line $y = 12 - x$.

**c** The line $y = 2x - 5$ is _____ to the line $y = 7 - \frac{1}{2}x$.

**d** The line $3y = 4 - x$ is _____ to the line $y = 3x + 2$.

**e** The line $\frac{x}{3} + \frac{y}{4} = 1$ is _____ to the line $\frac{x}{6} + \frac{y}{8} = 1$.

**4 a** Copy and complete the table of values for the graph
$y = x^2 - 5x + 4$ for $-1 \le x \le 6$.

| $x$ | −1 | 0 | 1 | 2 | 3 | 4 | 5 | 6 |
|---|---|---|---|---|---|---|---|---|
| $x^2$ | | | | 4 | | | | |
| −5x | | | | −10 | | | | |
| +4 | 4 | 4 | 4 | 4 | 4 | 4 | 4 | 4 |
| y | | | | −2 | | | | |

**b** Hence, plot the graph of $y = x^2 - 5x + 4$ for $-1 \le x \le 6$.

**c** Use your curve to estimate the minimum point of $y = x^2 - 5x + 4$.

# 3D shapes

1   For this solid, draw

   **a**   a plan of the shape

   **b**   a front elevation

   **c**   a side elevation.

2   These diagrams show the plan and front elevation of a solid.

   The arrow shows the direction from which the front
   elevation was drawn. Sketch the solid.

3   Draw a sketch diagram of each of these solids and state
   whether the solid is a prism or not a prism.

   **a**   cylinder            **b**   cuboid

   **c**   tetrahedron         **d**   square-based pyramid

4   Sketch a net for this cylinder. Calculate the total area
   of its faces giving your answer to 3 significant figures.

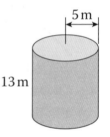

13 m   5 m

5   Draw a net for each of these solids, labelling the diagrams clearly.

   **a**

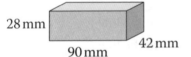

   28 mm   90 mm   42 mm

   **b**

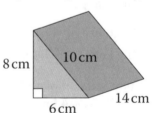

   8 cm   10 cm   6 cm   14 cm

# 5.2 Volume of a prism

**1** Work out the volume of the solids.

**a**

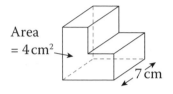

Area = 4 cm²

7 cm

**b**

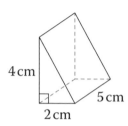

4 cm

2 cm

5 cm

**c**

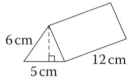

6 cm

5 cm

12 cm

**d**

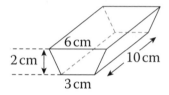

6 cm

2 cm

3 cm

10 cm

**2** The diagram shows a solid cylinder with a height of 12 cm and radius 5 cm.

Calculate the volume of the cylinder. Give your answer correct to 3 significant figures.

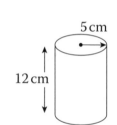

5 cm

12 cm

**3** A jeweller melts down a large bar of gold in the shape of the triangular prism shown to make smaller gold bars.

The bars are cuboids with dimensions 8 cm, 8 cm and 20 cm.

How many gold bars can she make?

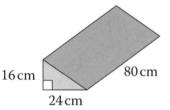

16 cm

24 cm

80 cm

**4** **a** Calculate the volume of a cube whose edges are all 5 cm.

**b** Calculate the volume of a cube whose faces have a total area of 384 cm².

**1** Find the volume of each solid.

**a** 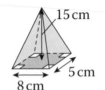 15 cm
5 cm
8 cm

**b** 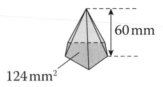 60 mm
124 mm²

**c**  4.2 m
12.6 m

**2** Find the surface area of each solid.

**a** Regular tetrahedron of side 4 cm.

**b** Square-based pyramid, base length = 10 cm, vertical height above mid-point of square base = 12 cm.

**3** Find the curved surface area of each cone.

**a** 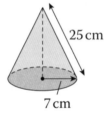 25 cm
7 cm

**b** 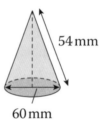 54 mm
60 mm

**c** 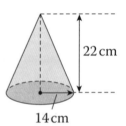 22 cm
14 cm

**4** Find **i** the volume and **ii** the surface area for each sphere.

**a** 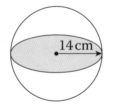 14 cm

**b** 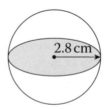 2.8 cm

**c** 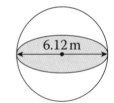 6.12 m

**5** A child's toy consists of a cone attached to a hemisphere. The radius of the hemisphere is 3 cm and the total height of the toy is 7 cm. Find

**a** the volume of the toy

**b** the surface area of the toy.

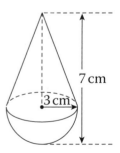 7 cm
3 cm

**6** A metalworker melts down a metal sphere of radius 8 cm to make 10 smaller spheres of equal volume.

Find the radius of a small sphere.

Q 1107, 1122, 1136   **SEARCH**

**1** These diagrams show the plan and front elevation of a solid.

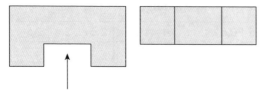

The arrow shows the direction from which the front elevation was drawn. Sketch the solid.

**2** Find the surface area and volume of this cylinder.
Give your answer to 3 significant figures.

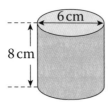

6 cm

8 cm

**3** Find the surface area of these solids.

**a**

35 mm

18 mm

**b**

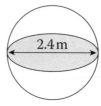

2.4 m

**c**

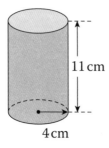

11 cm

4 cm

**d**

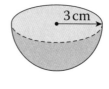

3 cm

**4** A sphere has a surface area of $100\,cm^2$.
Calculate the volume of the sphere.

# 16.1 Frequency diagrams

**1** The frequency table shows the time spent completing homework by a sample of students one Monday evening.

| Time, t (hours) | $0 < t \leq 0.5$ | $0.5 < t \leq 1$ | $1 < t \leq 2$ | $2 < t \leq 2.5$ | $2.5 < t \leq 3$ |
|---|---|---|---|---|---|
| Frequency | 56 | 32 | 32 | 10 | 6 |

   **a** Complete a histogram for the data.

   **b** How many students spent half an hour or less on their homework?

   **c** How many students were in the sample?

**2** This incomplete table and histogram give information about the length of 100 babies at birth.

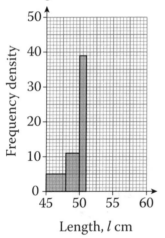

| Length, l (cm) | Frequency |
|---|---|
| $45 < l \leq 48$ | |
| $48 < l \leq 50$ | |
| $50 < l \leq 51$ | |
| $51 < l \leq 54$ | 18 |
| $54 < l \leq 56$ | 6 |

   **a** Use the histogram to complete the table.

   **b** Copy and complete the histogram.

**3** Work out the height of the second bar in each of these histograms in terms of $f$.

   **a**

| Length, l (cm) | Frequency | Bar height |
|---|---|---|
| $0 < l \leq 20$ | 16 | 4 cm |
| $20 < l \leq 60$ | $f$ | |

   **b**

| Time, t (min) | Frequency | Bar height |
|---|---|---|
| $0 < t \leq 15$ | 10 | 3 cm |
| $15 < t \leq 60$ | $f$ | |

**Averages and spread 2**

**1** Kath recorded the masses of 50 two-year-olds.

| Mass, _m_ (kg) | Frequency | Midpoint | Midpoint × frequency |
|---|---|---|---|
| $10 < m \leq 11$ | 5 | 10.5 | $10.5 \times 5 = 52.5$ |
| $11 < m \leq 12$ | 15 | 11.5 | |
| $12 < m \leq 13$ | 17 | | |
| $13 < m \leq 14$ | 9 | | |
| $14 < m \leq 15$ | 4 | | |
| **Total** | | | |

**a** Copy and complete the table to find an estimate for the mean.

**b** Find

**i** the modal class  **ii** the class containing the median.

**c** Work out an estimate for the range.

**2** The table shows the number of words in each of the first 50 sentences of the novel _Pride and Prejudice_ by Jane Austen.

**a** Calculate an estimate for the mean number of words per sentence.

**b** Is the mean a good representation of the average number of words per sentence? Explain your answer.

| No. of words per sentence | Frequency |
|---|---|
| 1–10 | 24 |
| 11–20 | 15 |
| 21–30 | 7 |
| 31–40 | 1 |
| 41–50 | 2 |
| 51–60 | 0 |
| 61–70 | 0 |
| 71–80 | 1 |

**3** This table shows the time taken in minutes by a group of 40 Mathematics teachers to solve a Sudoku puzzle.

| Time, _t_ (min) | $0 \leq t < 5$ | $5 \leq t < 10$ | $10 \leq t < 15$ | $15 \leq t < 20$ | $20 \leq t < 25$ |
|---|---|---|---|---|---|
| **Frequency** | 2 | 6 | 12 | 7 | 3 |

Use this information to find

**a** the modal class

**b** the class containing the median time

**c** an estimate of the mean time.

# 16.3 Box plots and cumulative frequency graphs

1 The following data give the heights, measured in centimetres, of a group of girls.

120, 131, 117, 108, 142, 137, 98, 116, 143
134, 128, 110, 139, 106, 126

a Work out the

   i   range          ii   median         iii   interquartile range

b Draw a box plot to show these data.

2 The following data were collected on the heights, measured in centimetres, of a group of boys.

lower quartile = 120    median = 127    maximum = 147

range = 39          interquartile range = 20

a Use this information to draw a box plot.

b Write down two comparisons between the boys' heights and the girls' heights in question **1**.

3 The table gives the exam results of a group of 120 students.

a Draw a cumulative frequency table and diagram for these data.

b Estimate (from the diagram)

   i   the median
   ii  the interquartile range.

| Test result, $t\%$ | Frequency |
|---|---|
| $40 < t \leq 50$ | 9 |
| $50 < t \leq 60$ | 18 |
| $60 < t \leq 70$ | 28 |
| $70 < t \leq 80$ | 44 |
| $80 < t \leq 90$ | 15 |
| $90 < t \leq 100$ | 6 |

c Estimate the number of students that passed the exam if the pass mark was 55%.

4 The cumulative frequency diagram shows the time taken for 50 girls to complete a puzzle.

The maximum time was 52 seconds and the minimum time was 9 seconds.

a Use the diagram to estimate the median time taken.

b Draw a box plot showing information about the girls' times.

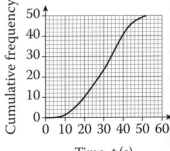

1194, 1195, 1333   SEARCH

# Scatter graphs and correlation

**1** The table gives the mean temperature (°C) in England and Wales for each month in one year and the number of ice cream cornets sold at Mr Frosty's ice cream parlour in Lincolnshire.

| Month | Jan | Feb | Mar | Apr | May | Jun |
|---|---|---|---|---|---|---|
| Temperature | 4.9 | 5.0 | 6.2 | 9.1 | 11.7 | 15.0 |
| Cornets | 15 | 18 | 20 | 52 | 115 | 164 |

| Month | Jul | Aug | Sep | Oct | Nov | Dec |
|---|---|---|---|---|---|---|
| Temperature | 15.4 | 17.1 | 14.4 | 10.3 | 7.5 | 5.3 |
| Cornets | 200 | 188 | 124 | 80 | 28 | 16 |

  **a** Represent these data on a scatter diagram.
  **b** Describe the correlation shown in terms of the data.

**2** The table gives the results of 10 students in two exam papers.

| Paper 1 | 48 | 65 | 85 | 96 | 75 | 59 | 69 | 84 | 53 | 90 |
|---|---|---|---|---|---|---|---|---|---|---|
| Paper 2 | 40 | 67 | 74 | 96 | 62 | 54 | 58 | 77 | 65 | 81 |

  **a** Represent these data on a scatter diagram.

  **b** Draw a line of best fit and use this to estimate a mark in paper 2 for a student who was absent on the day of the exam but who scored 73 on paper 1.

**3** In 1929 Edwin Hubble published a study of the relationship between how far away a galaxy is, measured in megaparsecs (Mpc), and how fast it appears to be moving away from us, measured in km/s. Here is some of his data.

| Distance (Mpc) | 0.03 | 0.28 | 0.62 | 0.68 | 0.91 | 1.05 | 1.4 | 1.63 | 2 |
|---|---|---|---|---|---|---|---|---|---|
| Speed (km/s) | 50 | 115 | 375 | 200 | 525 | 740 | 750 | 710 | 900 |

  **a** **i** Plot the data on a scatter graph.
     **ii** Describe any correlation you see.

  **b** **i** Draw a line of best fit.
     **ii** The Southern Pinwheel Galaxy was measured to be 0.9 Mpc away. Predict its speed away from us.
     **iii** The galaxy Messier 106 was measured to be moving away from us at 500 km/s. Predict its distance.

  **c** The quasar 3C 273 has been measured to be moving away from us at 47 000 km/s. Can you predict its distance? Explain your answer.

# 16.5 Time series

**1** A DIY store sells garden furniture. The table shows the number of units sold per month in one year.

| J | F | M | A | M | J | J | A | S | O | N | D |
|---|---|---|---|---|---|---|---|---|---|---|---|
| 1 | 0 | 2 | 6 | 18 | 30 | 16 | 14 | 5 | 15 | 3 | 2 |

  **a** Plot this data on a graph.

  **b** Describe any patterns in the data.

  **c** Suggest an explanation for the data for October.

**2** The following data show three years of ice cream sales, in pounds, at Miranda's delicatessen.

|  | Spring | Summer | Autumn | Winter |
|---|---|---|---|---|
| 2006 | 325 | 1850 | 625 | 900 |
| 2007 | 300 | 1750 | 650 | 875 |
| 2008 | 250 | 1700 | 600 | 825 |

  **a** Plot this data on a graph.

  **b** Describe any patterns in the data.

**3** The following data show the brightness of the star Delta Cephei measured over a number of days.

| Time (hours) | 0 | 16 | 48 | 64 | 92 | 100 | 112 |
|---|---|---|---|---|---|---|---|
| Brightness | 4.32 | 4.19 | 3.95 | 3.85 | 3.61 | 3.65 | 3.85 |
| Time (hours) | 120 | 130 | 160 | 200 | 215 | 245 | 260 |
| Brightness | 4.10 | 4.30 | 4.05 | 3.75 | 3.60 | 3.98 | 4.29 |

  **a** Plot this data on a graph.

  **b** Comment on any patterns in the data.

**4** The table shows the temperature of a patient after an operation recorded 10 times over 9 hours.

| Time | 1200 | 1300 | 1400 | 1500 | 1600 | 1700 | 1800 | 1900 | 2000 | 2100 |
|---|---|---|---|---|---|---|---|---|---|---|
| Temp (°C) | 39 | 39.7 | 40.4 | 41.6 | 42 | 40.8 | 40.3 | 39.7 | 38.4 | 37.1 |

  **a** Draw a time series graph to illustrate these figures.

  **b** Estimate the patient's temperature at

    **i** 1330      **ii** 1745      **iii** 2020.

Q 1198    SEARCH

**1** The table shows the number of words in each of the first 50 sentences of the novel *The Hobbit* by J. R. R. Tolkien.

| No. of words per sentence | Frequency |
| --- | --- |
| 1–10 | 13 |
| 11–20 | 10 |
| 21–30 | 9 |
| 31–40 | 7 |
| 41–50 | 8 |
| 51–60 | 2 |
| 61–70 | 1 |

   **a** Calculate an estimate for the mean number of words per sentence.

   **b** Write the modal class.

   **c** Write the class interval that contains the median.

**2** Harry summarised the heights of the large number of 'Queen Elizabeth' rose bushes planted in his garden.

Draw a box plot for the data.

> Smallest: 94 cm    Lower quartile: 106 cm    Median: 118 cm
>
> Tallest: 135 cm    Upper quartile: 124 cm

**3** A teacher marks a class set of books and records the number of minutes that each individual book takes him.

| No. of min, $m$ | $0 < m \le 1$ | $1 < m \le 2$ | $2 < m \le 3$ | $3 < m \le 4$ | $4 < m \le 5$ |
| --- | --- | --- | --- | --- | --- |
| Frequency | 1 | 6 | 10 | 7 | 6 |

   **a** Draw a cumulative frequency table for these data.
   **b** Draw a cumulative frequency diagram for these data.
   **c** Estimate (from the graph)
      **i** the median      **ii** the interquartile range.

**4** Danielle chose a random sample of 12 people and recorded their shoe size and their height in centimetres.

| Shoe size | 5 | 6 | 9 | 12 | 5 | 5 |
| --- | --- | --- | --- | --- | --- | --- |
| Height | 155 | 170 | 180 | 188 | 156 | 160 |
| Shoe size | 8 | 6 | 11 | 8 | 10 | 4 |
| Height | 168 | 165 | 183 | 172 | 180 | 152 |

   **a** Represent these data on a scatter diagram.
   **b** Describe the correlation shown in terms of the data.
   **c** Draw a line of best fit and use this to estimate the height of a person who has a shoe size of 7.

## 17.1 Calculating with roots and indices

**1** Simplify these expressions, giving your answer in index form.

   **a** $3^4 \times 3^5$      **b** $2^5 \times 2^2$      **c** $a^4 \div a$

   **d** $5^6 \div 5^6$      **e** $4^2 \times 4^2 \times 4^2$      **f** $x^2 \times x^3 \times x^4$

   **g** $6^5 \div 6^3 \times 6$      **h** $8^2 \times 8^6 \div 8^3$

**2** Solve these equations.

   **a** $2^x = 16$    **b** $y^2 = 5^4$    **c** $64 = (2^a)^2$    **d** $72 = b^3 \times 3^b$

**3** Evaluate these expressions.

   **a** $100^0$      **b** $500^1$

   **c** $(2^2)^4$      **d** $\left(\sqrt{7}\right)^2$

   **e** $36^{\frac{1}{2}}$      **f** $5^{-1}$

   **g** $8^{-\frac{1}{3}}$      **h** $4^{\frac{3}{2}}$

**4** Simplify these expressions, giving your answer in index form.

   **a** $(3^2)^4$    **b** $(5^3)^2$    **c** $\left(9^{\frac{1}{2}}\right)^4$    **d** $(8^4)^{\frac{1}{3}}$

**5** Evaluate these expressions.

   **a** $12^0$      **b** $25^{\frac{1}{2}}$      **c** $3^{-1}$

   **d** $27^{\frac{1}{3}}$      **e** $2^{-3}$      **f** $9^{-2}$

   **g** $36^{-\frac{1}{2}}$      **h** $16^{-\frac{1}{4}}$      **i** $8^{\frac{2}{3}}$

   **j** $16^{\frac{3}{4}}$

**6** Simplify these expressions, giving your answer as a power of 10.

   **a** $10^4 \times 10^8$      **b** $10^{12} \div 10^7$

   **c** $10^3 \div 10^9$      **d** $10^5 \times 10^4 \div 10^2$

   **e** $10^8 \div 10^4 \times 10^{-3}$      **f** $10^5 \div 10^{-4} \times 10^6$

   **g** $10^2 \times 10^2 \times 10^2 \times 10^2$      **h** $\dfrac{10^{-8} \times 10^3}{10^7 \times 10^{-9}}$

1033, 1045, 1301, 1924   **SEARCH**

# Exact calculations

**1** Simplify these expressions.

**a** $\sqrt{3} + \sqrt{3}$                   **b** $\sqrt{5} \times \sqrt{5}$

**c** $4\sqrt{2} + \sqrt{2}$                 **d** $\sqrt{20}$

**e** $(4 + \sqrt{7})(4 - \sqrt{7})$         **f** $(5 + \sqrt{2})(5 + \sqrt{2})$

**2** Simplify these expressions.

**a** $2\sqrt{5} + \sqrt{45}$      **b** $\sqrt{20}\,(1 + \sqrt{5})$      **c** $(5 + \sqrt{3})(5 - \sqrt{3})$

**3** Calculate these, leaving $\pi$ in your answers.

**a** The area of a circle of diameter 14 cm.

**b** The volume of a hemisphere of radius 6 m.

**c** The radius of a circle of circumference 18 mm.

**4** Rationalise the denominator of each of these fractions.

**a** $\dfrac{1}{\sqrt{5}}$                **b** $\dfrac{2}{\sqrt{7}}$

**c** $\dfrac{3}{1 + \sqrt{10}}$           **d** $\dfrac{1 - \sqrt{3}}{1 + \sqrt{3}}$

**5** Solve these equations using the method of completing the square, leaving your answers in surd form.

**a** $x + \dfrac{23}{x} = 10$      **b** $x + 6 + \dfrac{2}{x} = 0$      **c** $\dfrac{2}{x} + \dfrac{1}{x+1} = 1$

> Hint: In parts **a** and **b** multiply through by $x$ to obtain a quadratic equation.

# 17.3 Standard form

1 Write these numbers in standard form.

   **a** 600        **b** 19 340        **c** 2 000 000

   **d** 15        **e** 17 504        **f** 718 300

2 Write these numbers in standard form.

   **a** 0.16        **b** 0.005 32        **c** 0.060 01

   **d** 0.04        **e** 0.000 000 7        **f** 0.004 321

3 Change these numbers in standard form to ordinary numbers.

   **a** $3.6 \times 10^3$        **b** $5.91 \times 10^{-5}$

   **c** $2.15 \times 10^{-1}$        **d** $9.009 \times 10^2$

4 Evaluate these calculations, giving your answer in standard form. Do not use a calculator.

   **a** $(3 \times 10^2) \times (3 \times 10^4)$        **b** $(2.4 \times 10^2) \div (2 \times 10^4)$

   **c** $(3.2 \times 10^{-4}) \times (3 \times 10^{-2})$        **d** $(9.6 \times 10^{-6}) \div (3.2 \times 10^4)$

5 Work these out without using a calculator, giving your answer in standard form.

   **a** $(8 \times 10^4) \div (4 \times 10^2)$        **b** $(9.6 \times 10^{-8}) \div (3 \times 10^{-5})$

   **c** $(6 \times 10^{-4}) \times (5 \times 10^9)$        **d** $(2.4 \times 10^3) \times (5 \times 10^4)$

   **e** $(3 \times 10^5) \div (6 \times 10^{-2})$

6 Evaluate these calculations, giving your answer in standard form.

   **a** $(1.7 \times 10^5) + (3.2 \times 10^5)$        **b** $(9.4 \times 10^3) + (3.6 \times 10^3)$

   **c** $(4.2 \times 10^4) + (6.5 \times 10^3)$        **d** $(8.6 \times 10^5) - (3.5 \times 10^4)$

7 The population of Sweden is approximately $9 \times 10^6$ people.

   **a** Write this number given in standard form as an ordinary number.

   Sweden has an area of approximately 450 000 km².

   **b** Write this number in standard form.

   **c** Work out the population density without using a calculator and give your answer in standard form.

Q 1049, 1050, 1051    SEARCH

**1** Evaluate these expressions.

   **a** $(3^2)^2$       **b** $(\sqrt{7})^2$       **c** $\sqrt{(2^2 \times 3^2)}$

**2** Evaluate each of these expressions.

   **a** $8^{\frac{2}{3}}$            **b** $16^{\frac{3}{2}}$

   **c** $100^{\frac{5}{2}}$          **d** $81^{-\frac{1}{4}}$

   **e** $64^{-\frac{1}{2}}$         **f** $25^{-\frac{3}{2}}$

**3** Evaluate these, leaving your answers in surd form.

   **a** $\sqrt{3}\,(2 - \sqrt{3})$       **b** $(8 + \sqrt{5})\,(3 - \sqrt{5})$

**4** Work out these calculations and give your answers in index form.

   **a** $10^3 \times 10^{-2} \times 10^5$    **b** $10^4 \times 10^3 \div 10^5$

   **c** $10^2 \times 10^{-3} \div 10^7$    **d** $10^{-4} \times 10^{-3} \div 10^5$

**5** **a** Write these numbers in standard form.

     **i** 300     **ii** 843     **iii** 436 000     **iv** 0.007 53

   **b** Write these numbers as ordinary numbers.

     **i** $3.6 \times 10^4$   **ii** $9.06 \times 10^2$   **iii** $3 \times 10^5$     **iv** $6.11 \times 10^{-3}$

**6** Work these out without using a calculator, giving your answer in standard form.

   **a** $(3 \times 10^5) \times (2 \times 10^7)$        **b** $(8 \times 10^{10}) \div (4 \times 10^3)$

   **c** $(5 \times 10^6) \times (3 \times 10^4)$        **d** $(1.6 \times 10^5) \div (2 \times 10^2)$

# 18.1 Cubic and reciprocal functions

**1** **a** Copy and complete the table of values for the graph
$y = x^3 - 5x^2 + 2x + 8$ for $-2 \leq x \leq 5$.

| x | −2 | −1 | 0 | 1 | 2 | 3 | 4 | 5 |
|---|---|---|---|---|---|---|---|---|
| x³ | | | | | | 27 | | |
| −5x² | | | | | | −45 | | |
| +2x | | | | | | 6 | | |
| +8 | 8 | 8 | 8 | 8 | 8 | 8 | 8 | 8 |
| y | | | | | | −4 | | |

**b** Hence, plot the graph of $y = x^3 - 5x^2 + 2x + 8$
for $-2 \leq x \leq 5$.

**c** Use your curve to estimate the coordinates of
the turning points of the graph.

Hint: The turning points are where the graph changes from having a positive gradient to a negative gradient or vice versa.

**2** This is a sketch of the graph $y = x^3 - x^2 - 6x$.

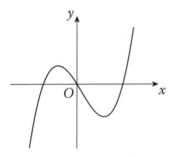

Hint: Fully factorise $x^3 - x^2 - 6x$. Take out a factor of $x$ to begin. Factorise the quadratic that remains.

Find the coordinates of the points where the
graph cuts the $x$-axis.

**3** **a** Copy and complete the table of values for the graph
$f(x) = \dfrac{6}{x}$ for $-6 \leq x \leq 6$.

| x | −6 | −5 | −4 | −3 | −2 | −1 | 0 | 1 | 2 | 3 | 4 | 5 | 6 |
|---|---|---|---|---|---|---|---|---|---|---|---|---|---|
| f(x) | | | | | | | | | | | 1.5 | | |

**b** Hence plot the graph of $f(x) = \dfrac{6}{x}$.

**c** Use your graph to estimate the value of $f(3.5)$.

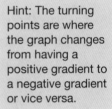

1071, 1172 SEARCH

# 8.2 Exponential and trigonometric functions

1  Using sketch graphs to help you as necessary

   a  list five values of $x$ for which $\sin x = 1$

   b  list five values of $x$ for which $\cos x = -1$.

2  Draw a graph of $y = \sin x$ for $0° \le x \le 360°$.
   On the same set of axes draw and label graphs of

   a  $y = 1 + \sin x$     b  $y = -\sin x$     c  $y = \sin(x + 90°)$

3  a  Plot the graph of $y = 3^{x-1}$ for $-2 \le x \le 4$.

   b  Use your graph to find an approximate value for $x$ when

      i  $y = 2$      ii  $y = 10$.

4  The graph shows the function $y = ab^x$.

   a  Given that the graph passes through
      $(1, 12)$ and $(3, 192)$, find $a$ and $b$.

   b  Explain why the curve will pass
      through the point $(4, 768)$.

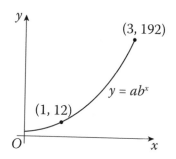

5  The graph of the function $f(x)$ is shown.
   What would the points $A$, $B$ and $C$ be
   translated to under the transformations

   a  $f(x) + 2$

   b  $f(x) - 3$

   c  $f(x - 2)$

   d  $f(x + 3)$?

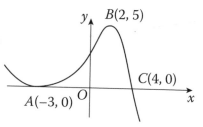

1   Which graph best represents each of these statements?

   **a**   The price of the US dollar against the pound had begun to rise but is now falling steadily.

   **b**   The cost of a company's shares has risen steadily during the past three months.

   **c**   The cost of petrol, which had been rising steadily in the last year, is now rising rapidly.

2   **a**   Plot a graph to represent the total cost of hiring a plumber, if their call-out charge is £50 and they charge £20 an hour labour. The longest that they will take is 8 hours.

   **b**   Use your graph to work out the cost of hiring a plumber for 4 hours.

   **c**   How long did they spend on a job if the bill was £160?

   **d**   Write the equation of the line, stating the meaning of any letters used.

3   This is a conversion graph to convert British pounds to Australian dollars.

   Use the graph to convert

   **a**   **i**   £50 to Australian dollars

      **ii**   AU$96 to pounds.

   **b**   If Julien has £60 and Alison has AU$120, who can purchase the most British goods if Alison converts her money?

   **c**   Write the equation of the line, stating the meaning of any letters used.

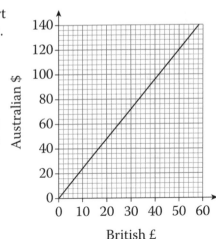

1184, 1322   **SEARCH**

# 3.4 Gradients and areas under graphs

**1** A sales company reimburses its staff for travel expenses.
This graph shows the amount reimbursed per km travelled.

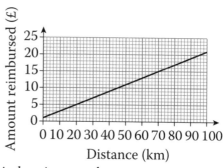

a Find the equation of the line, stating the meaning of any letters used.

b Interpret the meaning of $m$ and $c$ and discuss the limitations of this scheme for reimbursing travel costs.

**2** Find the gradients of the line segments between

   **i**   (1, 3) and (2, 6)    **ii**   (−1, 3) and (4, 8)    **iii**   (2, 5) and (5, −3)

   **iv**  (−6, 7) and (−2, −6) **v**   (−8, −5) and (3, 7)   **vi**   (−1, −3) and (2, −1)

**3** The diagram shows the graph of $y = x^2$ for values of $x$ from −4 to 4. $A$ is the point (1, 1) and $B$ is a point on the curve to the right of $A$.

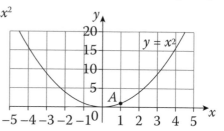

a Copy and complete the table below. The first line is done for you.

| Point B | | Gradient of chord AB |
| --- | --- | --- |
| $x$-coordinate | $y$-coordinate | |
| 4 | 16 | $(16 - 1) \div (4 - 1) = 5$ |
| 3 | | |
| 2 | | |
| 1.5 | | |
| 1.1 | | |
| 1.01 | | |

b Using these values, estimate the gradient of the curve at point $A$. Explain your answer.

**4** a Sketch the curve $y = (x + 1)^2$ for values of $x$ from 0 to 4.

b By dividing the curve into 4 vertical trapezia, estimate the area between the curve and the $x$-axis from $x = 0$ to $x = 4$.

## 18.5    Equation of a circle

1   By drawing appropriate graphs, find approximate solutions to the pair of simultaneous equations

$$x^2 + y^2 = 9$$

$$y = 2$$

> Hint: The equation of a circle of radius $r$ and centre $(0,0)$ is given by $x^2 + y^2 = r^2$.

2   **a**   By drawing appropriate graphs, find approximate solutions to the pair of simultaneous equations

$$x^2 + y^2 = 25$$

$$y = 2x + 1$$

    **b**   Confirm these solutions by solving the simultaneous equations algebraically.

3   **a**   Write down the equation of the circle with centre the origin and radius

      **i**   3     **ii**   9     **iii**   27     **iv**   7.5     **v**   12.4.

    **b**   For each of these circles write down the centre and radius.

      **i**   $x^2 + y^2 = 36$     **ii**   $x^2 + y^2 = 0.25$     **iii**   $x^2 + y^2 = 1000$

      **iv**   $x^2 + y^2 - 81 = 0$   **v**   $x^2 + y^2 + 64 = 0$

4   **a**   Do the points $(6, 7)$ and $(5, 12)$ lie on the circle

$$x^2 + y^2 - 169 = 0?$$

    **b**   Is the point $(5, 8)$ inside or outside the circle with centre the origin and radius 9.5?

5   **a**   Find the equation of the chord of the circle $x^2 + y^2 = 625$ between the points $(20, 15)$ and $(-7, 24)$.

    **b**   Find the equations of the tangents to the circle at these points.

**1** **a** Copy and complete this table to generate coordinates for the graph of $y = x^3 + 2x^2$.

| x | −3 | −2 | −1 | 0 | 1 | 2 |
|---|---|---|---|---|---|---|
| $x^3$ | −27 | | | | | |
| $2x^2$ | 18 | | | | | |
| y | −9 | | | | | |

**b** Draw and label an $x$-axis from −3 to 2 and a $y$-axis from −10 to 20. Plot the coordinates from the table in part **a**.

**c** Write the coordinates of the turning points of this cubic graph.

**2** Plot the graphs of these functions. Use an appropriate set of axes.

**a** $y = 3^x$      **b** $y = \cos x$      **c** $y = (x - 3)^2$

**3** The graph shows the function $y = f(x)$.
In terms of $f(x)$, write the equations of the transformed functions shown in the following graphs.

**a**

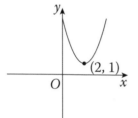

**b**

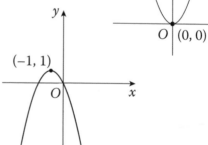

**4** Rain falls into these barrels at a steady rate. The graphs show how the water level changes. Match each graph to a barrel.

**a**      **b**      **c**

**i**      **ii**      **iii**

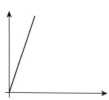 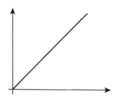 

# 19.1   Pythagoras' theorem

1   Calculate the lengths marked by letters.
Give your answers to 3 significant figures.

**a**
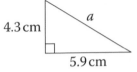
4.3 cm   *a*   5.9 cm

**b**
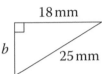
18 mm   *b*   25 mm

2   Calculate the lengths marked by letters.

**a**
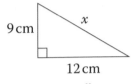
9 cm   *x*   12 cm

**b**
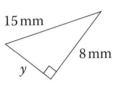
15 mm   8 mm   *y*

**c**
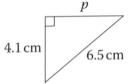
*p*   4.1 cm   6.5 cm

**d**
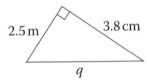
2.5 m   3.8 cm   *q*

3   Find the sides marked by letters in these diagrams.

**a**
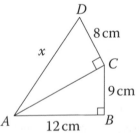
*D*   8 cm   *x*   *C*   9 cm   *A*   12 cm   *B*

**b**
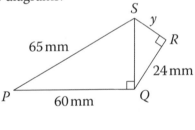
*S*   *y*   *R*   65 mm   24 mm   *P*   60 mm   *Q*

4   Use Pythagoras' theorem to work out the length of the line segment
joining these points.
   **a**   (1, 4) and (2, 8)   **b**   (2, −1) and (5, 5)   **c**   (−3, −2) and (−1, 3)

   Hint: Sketch out each line.

5   **a**   Work out the *exact* length of the diagonal of a square of side 4 cm.
   **b**   Work out the length of the diagonal of a rectangle with
      dimensions 7.5 cm and 3.5 cm.

6   Prove that this triangle is right-angled.
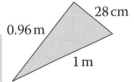
28 cm   0.96 m   1 m

   Hint: The converse of Pythagoras' theorem is true; that
   is, if the sides of the triangle are *a*, *b* and *c* such that
   $a^2 + b^2 = c^2$ then the triangle must be right-angled.

🔍 1064, 1112   **SEARCH**

# Trigonometry 1

**1** Find the angles marked by letters in each of these right-angled triangles, giving your answers to 3 significant figures.

**a**

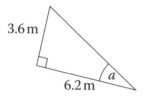

3.6 m
6.2 m
*a*

**b**

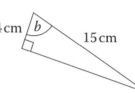

4 cm /*b*
15 cm

**c**

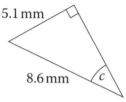

5.1 mm
8.6 mm
*c*

**d**

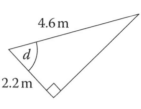

4.6 m
*d*
2.2 m

**e**

85 cm
45 cm
*e*

**f**

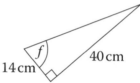

*f*
40 cm
14 cm

**2** Find the sides marked by letters in each of these right-angled triangles, giving your answers to 3 significant figures.

**a**

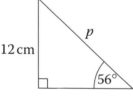

*p*
12 cm
56°

**b**

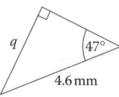

*q*
47°
4.6 mm

**c**

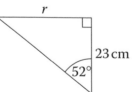

*r*
23 cm
52°

**d**

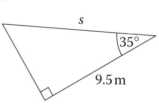

*s*
35°
9.5 m

**3** Using trigonometry, find *exact* values for these ratios.

    **a** $\sin 45°$     **b** $\cos 45°$     **c** $\tan 45°$

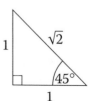

1
$\sqrt{2}$
45°
1

**1** Use the cosine rule to find the sides marked with letters.

**a**

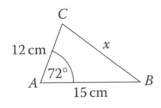

**b**

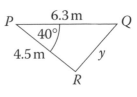

**2** Use the cosine rule to find the angles marked $\theta$.

**a**

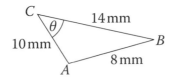

**b**

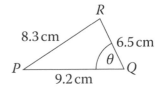

**3** Use the sine rule to find $AB$.

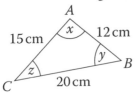

**4** In triangle $PQR$, angle $QPR = 54°$, angle $PQR = 32°$ and $PR = 10$ mm.
Find the two unknown sides of the triangle.

Hint: Find angle $QRP$.

**5** In triangle $ABC$, angle $BAC = 63°$, $AB = 3.3$ cm and $BC = 4.1$ cm.
Find the two unknown angles of the triangle.

**6** Find all three angles of triangle $ABC$.

**7** Quadville is 45 km from Parallelford on a bearing of 124°. Rhombustown
is 60 km due south of Parallelford.
By sketching a diagram and using the cosine rule, find the distance from
Rhombustown to Quadville.

Q 1095, 1120  **SEARCH**

# 9.4 Pythagoras and trigonometry problems

**1** The cross-section of this prism is an isosceles triangle with a base of 10 cm and a *slant* height of 13 cm.
The prism has a length of 30 cm.
Work out the volume of the prism.

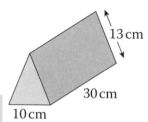

Hint: Use Pythagoras' theorem to find the perpendicular height of the triangular cross-section.

**2** A square has diagonals 20 cm long.
Use a sketch diagram to help you
  **a** work out the area of the square
  **b** work out the length of a side of the square.

Hint: Look at the square as two triangles.

Hint: Use square roots.

**3** The base *ABCD* of a cuboid is 8 cm by 15 cm.

The diagonal *AG* makes an angle 37° with the base of the cuboid.

What is the volume of the cuboid?

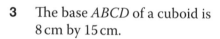

**4** A mountain is 1200 m high.
Due south of the mountain is Alphaville, from which the angle of elevation of the summit is 23°.
Due east of the mountain is Betatown, from which the angle of elevation of the summit is 35°.
What is the distance between Alphaville and Betatown?

**5** **a** Find the height of *XY*.
  **b** Find the angle *XAY*.

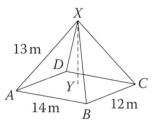

1 Draw these vectors on square grid paper.

$$a = \begin{pmatrix} -3 \\ 2 \end{pmatrix} \qquad b = \begin{pmatrix} 6 \\ -4 \end{pmatrix} \qquad c = \begin{pmatrix} -6 \\ -4 \end{pmatrix} \qquad d = \begin{pmatrix} 2 \\ 3 \end{pmatrix}$$

    **a** Write a pair of vectors that are parallel.

    **b** Write a pair of vectors that are perpendicular.

2 Use Pythagoras' theorem to find the magnitude of each of the vectors in question **1**.

3 *OABC* is a parallelogram.

    $\overrightarrow{OA} = \mathbf{a}$ and $\overrightarrow{OC} = \mathbf{c}$

    Find, in terms of **a** and **c**

    **a** $\overrightarrow{BC}$      **b** $\overrightarrow{AB}$      **c** $\overrightarrow{AC}$

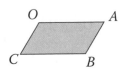

4 *OABC* is a trapezium.

    *OC* and *AB* are parallel.

    $AB = \frac{2}{3} OC$

    $\overrightarrow{OA} = \mathbf{a}$ and $\overrightarrow{OC} = \mathbf{c}$

    Find, in terms of **a** and **c**

    **a** $\overrightarrow{AB}$      **b** $\overrightarrow{OB}$      **c** $\overrightarrow{CB}$

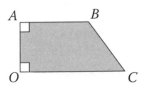

5 *ABCDEF* is a regular hexagon with centre *O*.

    $\overrightarrow{OA} = 2\mathbf{a}$ and $\overrightarrow{OB} = 2\mathbf{b}$

    **a** Express in terms of **a** and **b**

       **i** $\overrightarrow{AB}$     **ii** $\overrightarrow{BC}$

    **b** *M* is the midpoint of *AF*. Express $\overrightarrow{EM}$ in terms of **a** and **b**.

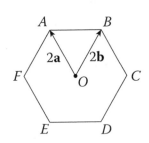

6 $\overrightarrow{OP} = \mathbf{p}$ and $\overrightarrow{OS} = \mathbf{s}$

    $\overrightarrow{PQ} = \frac{1}{2}\mathbf{p}$ and $\overrightarrow{SR} = \frac{1}{2}\mathbf{s}$

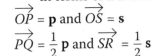

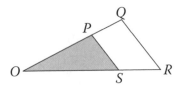

Prove that *PS* and *QR* are parallel.

**1** Work out the unknown sides in these right-angled triangles.

**a**

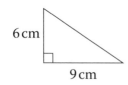

6 cm

9 cm

**b**

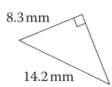

8.3 mm

14.2 mm

**c**

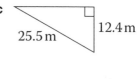

25.5 m

12.4 m

**2** Find the angles marked by letters in each of these right-angled triangles, giving your answers to 3 significant figures.

**a**

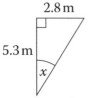

2.8 m

5.3 m

$x$

**b**

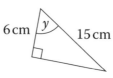

6 cm $y$ 15 cm

**c**

4.9 mm

$z$

7.3 mm

**d**

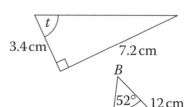

$t$

3.4 cm

7.2 cm

**3** In triangle $ABC$, angle $BAC = 84°$, angle $ABC = 52°$ and $BC = 12$ cm. Find the two unknown sides of the triangle.

Hint: Find angle $BCA$ and use the sine rule.

**4** In the cuboid, $PQ = 7$ m, $QR = 3$ m and $RS = 4$ m.

**a** Find $PS$.

**b** Find the angle $PS$ makes with the base of the cuboid.

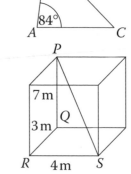

$B$

$52°$ 12 cm

$84°$

$A$ $C$

$P$

7 m

3 m $Q$

$R$ 4 m $S$

**5** $OPQR$ is a trapezium. $PQ$ and $OR$ are parallel. $PQ = 3OR$ $\overrightarrow{OP} = \mathbf{p}$ and $\overrightarrow{OR} = \mathbf{r}$ Find, in terms of $\mathbf{p}$ and $\mathbf{r}$

**a** $\overrightarrow{PQ}$  **b** $\overrightarrow{OQ}$  **c** $\overrightarrow{QR}$.

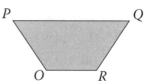

$P$ $Q$

$O$ $R$

# 20.1    Sets

**1**   **a**   List all the elements of these sets.

     **i**   $C \equiv$ first 5 cube numbers    **ii**   $T \equiv$ first 10 triangular numbers

     **iii**   $F \equiv$ first 10 terms in the Fibonacci sequence starting 1, 1, ...

     **iv**   $M \equiv$ multiples of 5 less than 100

  **b**   Using the sets in part **a**, write the elements of

     **i**   $C \cup T$       **ii**   $C \cap T$       **iii**   $T \cap F$

     **iv**   $F \cap M$      **v**   $C \cap T \cap F$     **vi**   $M \cap T \cap F$

**2**   Give precise descriptions of these sets.

  **a**   {lunes, martes, miércoles, jueves, viernes, sábado, domingo}

  **b**   {1, 2, 3, 4, 6, 8, 12, 24}      **c**   {6, 28, 496, 8128}

  **d**   {0, 3, 8, 15, 24, 35, 48, 63, 80, 99}

**3**   Draw Venn diagrams to represent these sets.

  **a**   Rectangles and squares

  **b**   Parallelograms and rhombuses

  **c**   **i**   $F \equiv$ factors of 48      **ii**   $M \equiv$ multiples of 3 less than 48

  **d**   **i**   $F \equiv$ factors of 48      **ii**   $S \equiv$ Square numbers less than 50

  **e**   $F \equiv$ factors of 48 $\cap$ $P \equiv$ powers of 2 less than 64

  **f**   $F \equiv$ prime numbers less than 30; $E \equiv$ even numbers less than 30

**4**   100 students are allowed to select up to two activities from badminton, golf and rugby.

  41 students opt to play badminton; 17 play golf only;

  5 play badminton and rugby; 20 play badminton and golf;

  27 play rugby only.

  **a**   Draw and complete a Venn diagram to show these data.

  **b**   How many students

     **i**   play all three sports      **ii**   play rugby and golf

     **iii**   play rugby            **iv**   play golf

     **v**   play rugby or golf      **vi**   play two sports

     **vii**   play one sport only?

Q 1262, 1921, 1922   **SEARCH**

# Possibility spaces

1 A spinner has equally sized sections numbered 1 to 4. A fair dice is numbered 1 to 6. A student spins the spinner and throws the dice. She adds the scores together.

   a Copy and complete the table to show all possible outcomes.

|  |  | Dice | | | | | |
|---|---|---|---|---|---|---|---|
|  |  | **1** | **2** | **3** | **4** | **5** | **6** |
| **Spinner** | **1** | 2 | 3 | 4 |  |  |  |
|  | **2** | 3 |  |  |  |  |  |
|  | **3** |  |  |  |  |  |  |
|  | **4** |  |  |  |  |  |  |

   b Find the probability that the sum of the scores is

      **i**   10    **ii**   7    **iii**   not 5    **iv**   2 or 3.

2 A spinner has equally sized sections numbered 1 to 8. David spins the spinner and throws a fair coin.

   a Draw a table to show all possible outcomes.
   b Find the probability he obtains
      **i**    the number 2 and a head
      **ii**   a number greater than 5 and a tail
      **iii**   an even number and a tail
      **iv**   a prime number and a head.

3 Wedding guests are given a choice of either 'Aberdeen Angus fillet steak with a whisky and cream sauce' or 'Roast pave of Scottish salmon with a mushroom sauce'. The table shows the choices of 100 guests.

|  | **Steak** | **Salmon** | **Total** |
|---|---|---|---|
| **Male** | 42 | 18 | 60 |
| **Female** | 16 | 24 | 40 |
| **Total** | 58 | 42 | 100 |

   a The photographer chooses one wedding guest at random. What is the probability that this guest is female?
   b One of the male guests is chosen at random. What is the probability that he chooses the steak?
   c One of the guests who chose the salmon is picked at random. What is the probability that they are female?

## 20.3 Tree diagrams

**1** An artist has a box of watercolour paints containing 4 pans of cadmium red colour and 6 pans of cobalt blue colour. In order to paint an abstract picture, the artist chooses a pan of colour at random from the box. He paints with this colour and then chooses a second pan of colour at random.

**a** Copy and complete the tree diagram to show all possible outcomes.

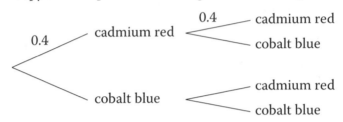

**b** Find the probability that the artist chooses cobalt blue both times.

**2** Bag A contains 7 white marbles and 3 black marbles.
Bag B contains 2 white marbles and 4 red marbles.
A student chooses a marble at random from each bag.
**a** Draw a tree diagram to show all possible outcomes.
**b** Find the probability that the two marbles chosen are different colours.
**c** Find the probability that at least one of the marbles chosen is red.

**3** A tin contains 12 coloured pencils: 8 yellow and 4 green.
A student chooses a pencil at random from the tin, uses it to create a drawing and then replaces it. She then picks a second pencil from the tin at random.
**a** Draw a tree diagram to show all possible outcomes.
**b** Find the probability that the two pencils chosen are
    **i** both yellow         **ii** one of each colour.
**c** Find the probability that at least one of the pencils chosen is green.

**4** A teacher requires two classroom monitors to be in charge of the register. He chooses two students at random from a class of 25 students: 15 boys and 10 girls.
**a** Draw a tree diagram to show all possible outcomes.
**b** Find the probability that the two students chosen are
    **i** both girls         **ii** a boy or a girl, in any order.
**c** Find the probability that at least one of the students chosen is a boy.

Q 1208, 1334, 1935    **SEARCH**

# Conditional probability

**1**   Jason and Clare play two games of tennis. The probability that Jason will win any game against Clare is 0.55. Work out the probability that Jason wins at least one game.

**2**   Lucy is offered a biscuit from each of two tins.
The first tin contains 6 chocolate and 4 plain biscuits.
The second tin contains 5 chocolate and 10 plain biscuits.
Lucy chooses one biscuit at random from each tin. Calculate the probability that the chosen biscuits are different types.

**3**   Isobel sits a multiple choice test with three questions. Each question has five possible answers. Isobel chooses the answer to each question at random. In order to pass the test, Isobel must correctly answer at least two of the three questions.

   **a**   What is the probability that Isobel will correctly answer a question?

   **b**   Draw a tree diagram to show all possible outcomes.

   **c**   Find the probability that Isobel passes the test.

**4**   The table shows information about the teachers in a school.

| | ≤5 years at school | >5 years at school |
|---|---|---|
| **Male** | 20 | 30 |
| **Female** | 15 | 10 |

   **a**   A headteacher chooses two teachers at random to accept an award for their school. He chooses one male and one female teacher. Calculate the probability that they will both have worked at the school for ≤5 years.

   **b**   Later he decides to choose one teacher with ≤5 years service and one with >5 years service to the school. Calculate the probability that both teachers are male.

**5**   The probability that Ava eats breakfast is 0.4. If Ava eats breakfast, the probability that she is late for school is 0.3 If Ava does not eat breakfast, the probability that she is late for school is 0.1.

   **a**   Draw a tree diagram to show all possible outcomes.

   **b**   Work out the probability that on any one day, Ava will *not* be late for school.

1 Ellie throws three coins together.

    **a** List all the possible outcomes of this experiment.

    **b** Find the probability that the three coins show

        **i** three heads

        **ii** a head and two tails in any order.

2 To start a game, a player must obtain a double-six on the throw of two fair dice.

    **a** When the player throws any one fair dice, what is the probability of *not* obtaining a six?

    **b** Draw a tree diagram to show the two events 'six' and 'not six' for each of the two fair dice.

    **c** Calculate the probability of obtaining at least one six.

    **d** Calculate the probability of obtaining a double-six.

3 A bag contains 4 red counters and 6 black counters.

Michael draws a counter at random, notes its colour and then replaces it in the bag. A second counter is then drawn.

    **a** Find the probability he draws

        **i** a red counter on the first draw followed by a black counter on the second

        **ii** two red counters.

    **b** What can you say about the events 'a red counter on the first draw' and 'a black counter on the second'?

    **c** If the first counter is not replaced before the second counter is drawn, what can you now say about the events 'a red counter on the first draw' and 'a black counter on the second'? Explain your answer.

Q 1166 **SEARCH**

# 1.1 Linear sequences

**1** Write the next two terms in each of these sequences.

   **a**   4, 8, 12, 16, 20, ...        **b**   3, 7, 11, 15, 19, ...

   **c**   100, 94, 88, 82, 76, ...     **d**   9, 16, 25, 36, 49, ...

   **e**   2, 4, 8, 16, 32, ...        **f**   1000, 500, 250, 125, 62.5, ...

   **g**   1, 1, 2, 3, 5, 8, ...        **h**   1, 2, 6, 24, 120, ...

**2** Generate the first five terms of each of these sequences.

   **a**   $T(n) = 4n - 3$      **b**   $T(n) = 50 - 3n$   **c**   $T(n) = n^2 + 5$

   **d**   $T(n) = (n + 1)(n + 3)$    **e**   $T(n) = \dfrac{n}{n+1}$     **f**   $T(n) = \dfrac{1}{n^2}$

**3** Find the $n$th term of these sequences.

   **a**   6, 11, 16, 21, 26, ...      **b**   20, 17, 14, 11, 8, ...

   **c**   $\dfrac{1}{7}, \dfrac{2}{11}, \dfrac{3}{15}, \dfrac{4}{19}, \dfrac{5}{23}, \ldots$     **d**   5, −1, −7, −13, −19, ...

**4** Find the $n$th term of these linear sequences.

   **a**   7, 9, 11, 13, 15, ...      **b**   3, 6, 9, 12, 15, ...

   **c**   3, 8, 13, 18, 23, ...      **d**   −8, −4, 0, 4, 8, ...

   **e**   17, 16, 15, 14, 13, ...    **f**   20, 17, 14, 11, 8, ...

   **g**   $1\dfrac{1}{2}, 2, 2\dfrac{1}{2}, 3, 3\dfrac{1}{2}, \ldots$     **h**   $4\dfrac{3}{4}, 4\dfrac{1}{2}, 4\dfrac{1}{4}, 4, 3\dfrac{3}{4}, \ldots$

**5** Given the following terms in a sequence

      15th term = 117

      16th term = 125

      17th term = 133

  find an expression for the $n$th term of the sequence.

**6** Find the value of $n$ that generates the term given from each of these sequences.

   **a**   $T(n) = 6n - 1$; $T(n) = 59$     **b**   $T(n) = (n - 1)^2$; $T(n) = 49$

   **c**   $T(n) = 5(n + 1)$; $T(n) = 75$    **d**   $T(n) = n^3 + 4$; $T(n) = 68$

> **Hint:** Form an equation by letting the $n$th term and the term itself equal one another.

## 21.2 Quadratic sequences

**1** Find the next two terms of these quadratic sequences.

   **a**   4, 8, 14, 22, ☐, ☐, ...        **b**   7, 9, 13, 19, ☐, ☐, ...

   **c**   2, 5, 10, 17, ☐, ☐, ...       **d**   10, 20, 26, 28, ☐, ☐, ...

**2** Find the missing terms in these quadratic sequences.

   **a**   8, 14, 24, ☐, 56, ...         **b**   ☐, 10, 19, 34, 53, ...

   **c**   1, 6, ☐, 31, ☐, ...           **d**   6, –8, –26, ☐, ☐, ...

**3**  **a**   Find the first three terms for the sequences described by these rules.

      **i**    $T(n) = n^2 + 6$         **ii**    $T(n) = n^2 + 3n$

      **iii**   $T(n) = n^2 + 2n - 6$      **iv**   $T(n) = 3n^2$

      **v**    $T(n) = 3n^2 + 5$

   **b**   Calculate the 10th term of each sequence.

**4**  **a**   Using the $n$th term for each sequence, calculate the first four terms.

      **i**    $n^2 + 4$            **ii**    $n^2 - 7$

      **iii**   $n^2 + n + 2$       **iv**   $n^2 + n - 9$

      **v**    $3n^2 + n$          **vi**   $n^2 + 0.5n$

      **vii**   $3n^2 + 4n + 5$      **viii**   $n(n - 5)$

   **b**   Calculate the second difference in each case to check the sequences are quadratic.

**5** Find the $n$th term of these quadratic sequences.

   **a**   6, 9, 14, 21, ...          **b**   –5, –2, 3, 10, ...

   **c**   6, 12, 22, 36, ...        **d**   –3, 1, 7, 15, ...

   **e**   3, 10, 19, 30, ...        **f**   13, 18, 25, 34, ...

**6** Find the $n$th term of these sequences.

   **a**   8, 5, 0, –7, ...            **b**   20, 14, 4, –10, ...

Q 1166    **SEARCH**

# Special sequences

**1** **a** **i** What is the first cube number above 1 to also be a square number?

  **ii** What is the first triangular number above 1 to also be a square number?

**b** Copy and complete the table.

| Term | 1 | 2 | 3 | 4 | 5 | $n$ |
|------|------|------|----|----|----|---------|
| **Sequence A** | 12 | 7 | | −3 | | $17 - 5n$ |
| **Sequence B** | 1 | | 27 | 64 | | |
| **Sequence C** | 1024 | 512 | | | 64 | |
| **Sequence D** | 3 | 8 | 15 | 24 | | |

**c** **i** What value of $n$ gives −483 in sequence A?

  **ii** What value of $n$ gives 125 000 in sequence B?

  **iii** What value of $n$ gives 1 in sequence C?

  **iv** What value of $n$ gives 3599 in sequence D?

**2** **a** A sequence begins −6, −3, 0, ...

Find the $n$th term of the sequence.

**b** A sequence has the terms: $T(1) = 1^2 - 3 = -2$; $T(2) = 2^2 - 6 = -2$; $T(3) = 3^2 - 9 = 0$.

Write down the next three terms and the $n$th term.

**c** Find the value of $n$ such that the ratio of the first sequence to the second sequence is $\frac{1}{3}$.

**3** **a** A sequence has a second term of 8 and a fourth term of 128.

  **i** If the sequence is arithmetic, find the common difference and the first term.

  **ii** If the sequence is geometric, find the common ratio and the first term.

**b** In an arithmetic sequence the tenth term is four times the fourth term. If the constant difference is $d$, and the first term is $a$, show that $d = -a$.

**1** Find the $n$th term of these sequences.

    **a**   4, 7, 10, 13, 16, ...        **b**   2, 7, 12, 17, 22, ...

    **c**   18, 14, 10, 6, 4, ...        **d**   1.5, 2, 2.5, 3, 3.5, ...

**2** Given the following terms of a sequence, find the $n$th term of the sequence.

    12th term = 67   13th term = 72   14th term = 77

**3** Describe the following sequences using one of these words.

> arithmetic geometric
> quadratic Fibonacci-type

    **a**   4, 7, 10, 13, ...        **b**   2, 4, 8, 16, ...

    **c**   4, 8, 12, 20, 32, ...        **d**   3, 7, 15, 27, ...

    **e**   4, −12, 36, −108, ...        **f**   −4, −7, −11, −16, ...

    **g**   −16, −13, −8, −1, ...        **h**   −3, −6, −9, −15, ...

**4** Find the next three terms of the following sequences using the properties of the sequence.

    **a**   arithmetic   2, 5, ☐, ☐, ☐, ...

    **b**   geometric   2, 5, ☐, ☐, ☐, ...

    **c**   Fibonacci   2, 5, ☐, ☐, ☐, ...

    **d**   quadratic   2, 5, ☐, ☐, ☐, ...

**5** **a** Find the first three terms for the sequences described by these rules.

    **i**   $T(n) = n^2 + 10$        **ii**   $T(n) = n^2 + 4n$

    **iii**   $T(n) = n^2 + n - 1$        **iv**   $T(n) = 4n^2$

    **v**   $T(n) = 5n^2 + 2$

    **b** Calculate the 10th term of each sequence.

**6** Find the $n$th term of these quadratic sequences.

    **a**   5, 9, 21, 41, ...        **b**   9, 7, 3, −3, ...

    **c**   1, 3.5, 8, 14.5, ...        **d**   2, 2, −6, −22, ...

# 2.1 Compound units

1 Caroline sets off from home for work at 7:45 a.m. and arrives at 8:10 a.m. She drives at an average speed of 56 km/h.

How far does Caroline travel to work?

2 A solid iron bar is in the shape of a cuboid of width 2 cm, height 12 cm and length 30 cm.

Iron has a density of 7.87 g/cm$^3$.

Work out the mass of the iron bar in kilograms to 3 significant figures.

3 A block of aluminium is in the shape of a cuboid of dimensions 3.4 m by 0.5 m by 2.1 m. Aluminium has a density of 2.7 g/cm$^3$. Work out the mass of the block of aluminium in kilograms.

4 The distance from Earth to the Sun is approximately $1.44 \times 10^8$ km.

   a Change this number from standard form to an ordinary number.

   b If light travels at a speed of approximately $3 \times 10^5$ km/s, work out how long it takes for light to travel from the Sun to Earth.

5 Blank DVDs are sold in two different packs.

A pack of 4 DVDs costs £4.99.

A pack of 5 DVDs costs £6.19.

Which is the better buy?

6 Melissa earns £8.70 per hour for 40 hours each week and 'time-and-a-half' if she works more than 40 hours. Melissa gets paid 'double time' if she works on Sundays. Calculate Melissa's pay if she works 45 hours from Monday to Friday and 8 hours on Sunday.

## 22.2 Converting between units

**1** Jamie's garden is twice as long as it is wide. The area of the garden is 162 m².

   **a** Write the area of the garden in square centimetres.

   **b** Calculate the length and width of the garden.

   **c** Jamie wants to construct a path to run along the diagonal of the garden. Calculate the length of this path.

**2** Two similar cylinders have diameters of 5 cm and 8 cm.

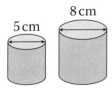

   If the capacity of the larger cylinder is 384 cm³, find the capacity of the smaller cylinder.

**3** The heights of two similar bottles of water are in the ratio 2 : 5. The smaller bottle has a capacity of 320 ml. What is the capacity of the larger bottle?

**4** Find the volume of this frustum, giving your answer to 3 significant figures.

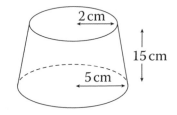

> Hint: First, use similar triangles to find the height of the complete cone. Consider the problem as the difference in volume between a large and small cone.

**5** The scale factor of a map is 1 : 20 000.

   **a** What length, in metres, does 1 cm on the map represent?

   **b** A plot of land is represented by 2 cm² on the map. How large, in square metres, is the actual plot of land?

**1**   The mass of gold, $g$, is directly proportional to its volume, $v$.

   **a**   Given that 200 cm³ of gold has a mass of 3864 g, find a formula connecting $g$ and $v$.

   **b**   Find the mass of a gold bar, in kilograms, with dimensions 20 cm by 8 cm by 4 cm.

   **c**   Find the volume of a gold bar with a mass of 46.368 kg. Suggest a reasonable set of dimensions for this gold bar.

**2**   $P$ varies with the square of $q$. If $P = 51.2$ when $q = 3.2$

   **a**   find a formula for $P$ in terms of $q$

   **b**   find the value of $P$ when $q = 0.3$

   **c**   find the value of $q$ when $P = 16.2$.

**3**   $y$ varies inversely with the square of $x$. If $y = 12.8$ when $x = 5$

   **a**   find a formula for $y$ in terms of $x$

   **b**   find the value of $y$ when $x = 2$

   **c**   find the value of $x$ when $y = 5$.

**4**   The variable $y$ is inversely proportional to the variable $x$.
   Write the effect on $y$ if $x$ is

   **a**   multiplied by 2        **b**   divided by 5

   **c**   multiplied by 0.25     **d**   divided by 0.8.

**5**   If $y$ is proportional to the square of $x$ and when $x = 3$, $y = 27$, calculate

   **a**   the value of $y$ when $x = 4$        **b**   the value of $x$ when $y = 75$.

**6**   If $a$ varies as $b$ and $a = 6$ when $b = 10$, find

   **a**   the value of $a$ when $b = 18$

   **b**   the value of $b$ when $a = 9$.

**7**   The number of people, $n$, required to build a wall is inversely proportional to the time taken, $t$.

   **a**   Given that it takes 10 people 4 hours to build this wall, find a formula connecting $t$ and $n$.

   **b**   How long would it take 8 people to build this wall?

   **c**   How many people are required to build the wall in just 20 minutes?

## 22.4 Rates of change

**1** Hicham el Guerrouj of Morocco holds the world record for both the 1500 metres and mile, with times of 3 m 26 s and 3 m 43 s (both to the nearest second). [1 mile $\equiv$ 1609.3 metres (1 dp)]

  **a** Find Hicham's average speed for both these records in

    **i** m/s                       **ii** km/h.

  **b** Usain Bolt currently holds the world 100 metres record at 9:58.

    **i** Find the ratio of Usain's speed to Hicham's speed for both distances.

    **ii** If Usain could maintain his 100 metres speed, how long would it take him to run the 1500 metres and mile distances?

**2** The 8030 kilometres long Trans-Canada Highway runs from Mile 0 at Victoria, British Columbia, to St John's, Newfoundland. The speed limit generally is 100 km/h. Debs and Dave drive the route, averaging 6 hours driving per day at 90% of the speed limit. How long does the journey take them?

**3** There are 5 sketch graphs of proportion below. Match each of the statements with the appropriate graph.

  **a**            **b**            **c**            **d**            **e**

    **i** The fall in water level as a conical funnel is filled at a constant rate.

    **ii** The fall in level as fluid drips from a cubical tank at a constant rate.

    **iii** The rise in water level as water is poured at a constant rate into a narrow necked vessel.

    **iv** The distance travelled by sphere rolling down an inclined groove.

    **v** The amount of water pouring from a hose pipe at a constant rate.

**1** Jake has £3500 in a bank account that earns 4.5% compound interest.

Rowan has £3500 in a bank account that earns 4.7% simple interest.

After 3 years, who will have the most money in their account assuming that no withdrawals have been made?

**2** A bank pays compound interest of 4%. Find the amount of money in an account after 3 years if the original investment is £5000.

**3** Philip's van depreciates in value by 8% each year. After four years the van is worth £9170.

   **a** What was the original cost of the van?

   **b** What was the overall percentage depreciation of the van after 4 years?

   **c** If the van continues to depreciate at the same rate, after how many years will the van be worth less than £5000?

**4** In 2011, the United Kingdom had a population of 62.2 million. If the population of the United Kingdom is increasing at, on average, an annual rate of 0.6%, calculate an estimate for the population of the United Kingdom in 2015. Give your answer to 3 significant figures.

**5** **a** The number of hits on a website is increasing in the ratio $5:4$ each month. Initially, the website received 10 000 hits. How many hits will it have after 3 months?

   **b** A balloon being blown up is expanding at the rate of $6:5$ each second. Originally, its volume was $5\,cm^3$. What is the volume after 10 seconds?

   **c** A wet carpet is drying at the rate of $2:5$ each hour. Originally the wet area covered an area of 2 square yards. How large is the wet area after 4 hours?

1   The mass, $M$, of copper, is directly proportional to its volume, $V$.

   a   Given that $150\,cm^3$ of copper has a mass of $1344\,g$,
       find the value of $p$ in the formula $M = pV$.

   b   What property of copper does the value $p$ represent?

   c   Find the mass of a solid copper rod of radius $1.5\,cm$ and length $30\,cm$.

   d   Find the volume of a solid copper rod with a mass of $2.912\,kg$.

2   The time taken, $t$(hours), on a journey varies inversely as the average
    speed, $s$ (km per hour), for the journey. When $t = 2.5$, $s = 48$.

   a   Write a formula for $t$ in terms of $s$.

   b   Calculate the value of $t$ when $s = 50$.

3   Mike wants to invest £200 for five years. His bank offers him two options.
    Option 1 is simple interest of 5.25% per annum.
    Option 2 is compound interest of 5% per annum.
    Which option should Mike choose in order to achieve the most interest
    on his investment? Show all your working.

4   a   Complete the following statements. The first is done for you.

   i   $4\,cm \equiv 40\,mm$   ii   $6\,m \equiv \_\,cm$   iii   $55\,mm \equiv \_\,cm$

   iv   $425\,m \equiv \_\,km$   v   $750\,ml \equiv \_$ litres   vi   $3.2\,kg \equiv \_\,g$

   vii   $3\,cm^2 \equiv \_\,mm^2$   viii   $1\,500\,000\,cm^2 \equiv \_\,m^2$   ix   $2.5\,m^3 \equiv \_\,cm^3$

   b   Answer the following questions, giving your answers in both units.

   i   $3\,mm + 4\,cm$   ii   $4\,km - 555\,m$

   iii   $3.2$ litres $+ 450\,ml$   iv   $2\,kg - 720\,g$

   c   i   A coach leaves Exeter bus station at $08:34$ and arrives at
           Heathrow airport at $11:54$, a distance of 200 miles. Calculate
           the average speed of the coach.

      ii   If one mile is equivalent to $1600\,m$, calculate the coach's average
           speed in km/h.

      iii   A map showing this journey has a scale of $1:500\,000$. How far
            apart are Exeter bus station and Heathrow airport on the map?
            Give your answer in inches and cm.

5   a   The number of people infected by a virus increases by 7.5% per
        week. How long will it take for the number of people infected to
        double in size?

   b   A vaccine reduces the number of patients infected with this virus
       at a rate of 15% per week. If $100\,000$ infected patients are given the
       vaccine, how many are still infected after 15 weeks?

# Revision homework

1 Use BIDMAS to work these out.

a $8 \times 9 + 5$

b $12 \div 4 \times 3$

c $12 \times (17 - 5)$

d $3 \times 5 + 6 \times 4$

e $8 \times (9 - 3) \times 2$

f $6^2 + 15 \div 3$

g $\dfrac{5 \times (4^2 - 2)}{7}$

h $\sqrt{120 - 2^3 \times 7}$

2 A student selects a letter, at random, from the letters of the word 'INDEPENDENCE'. Find the probability that the letter chosen is

a a D

b a consonant

c not an E

d from the second half of the alphabet.

3 Describe fully the transformation that maps the shaded rectangle onto the *un*shaded rectangle.

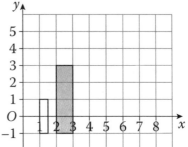

4 $x$ appears twice in each of these formulae. Collect the terms in $x$ on one side and rearrange to make $x$ the subject of the formula.

a $px + q = ax + b$

b $k(x - m) = tx$

c $\dfrac{x + p}{x - p} = A$

d $x(m - n) = p - x$

5 Calculate the interior angle sum of a hexagon, using diagrams to show your working.

6 Here are the times, in minutes, taken to complete some crossword puzzles.

| 12 | 15 | 21 | 10 | 32 | 9 | 18 | 24 | 20 | 8 |
|----|----|----|----|----|----|----|----|----|----|
| 26 | 22 | 19 | 11 | 28 | 23 | 23 | 14 | 25 | 27 |

Find the median and interquartile range of these data and comment on any trend.

**2** # Revision homework

1 A, B and C are points on the circumference of a circle, centre O.

   **a** Find angle AOC.

   **b** Give a reason for your answer to part **a**.

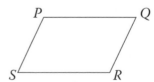

2 Expand and simplify.

   **a** $5(3 - 4x)$       **b** $6a(2a + 3b)$

   **c** $y^2(y + x)$       **d** $(x + 3)(x + 7)$

   **e** $(2t - 3)(t + 5)$    **f** $(p + 7)^2 + 2(p - 3)$

3 Find the upper and lower bounds of these calculations.

   **a** The area of a sheet of paper with length 29.4 cm and width 20.7 cm, measured to the nearest millimetre.

   **b** The speed of a car travelling 164 km, to the nearest kilometre, in 120 minutes, to the nearest minute.

4 The probability of Zahir being early or on time for work is 0.76.

   **a** Calculate the probability of his being late for work.

   **b** Over 300 work days, how many times would you expect Zahir to be late for work?

5 Write a formula for the area of each shaded region.

   **a**

   **b**
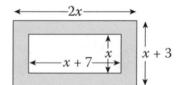

6 PQRS is a parallelogram. Prove that triangles SPQ and QRS are congruent.

# Revision homework

**1** Work out these calculations.

**a** $\dfrac{3}{4} \times \dfrac{1}{8}$    **b** $\dfrac{3}{5} \times \dfrac{5}{9}$    **c** $\dfrac{5}{8} \div \dfrac{2}{3}$    **d** $\dfrac{5}{12} \div \dfrac{1}{6}$

**e** $\dfrac{7}{15} \times 12$    **f** $10 \div \dfrac{2}{3}$    **g** $2\dfrac{2}{5} \times 3\dfrac{1}{3}$    **h** $4\dfrac{2}{7} \div 3\dfrac{1}{8}$

**2** These two parallelograms are mathematically similar.

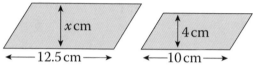

$x$ cm

12.5 cm

4 cm

10 cm

Find the missing value $x$ and hence the area of the larger parallelogram.

**3** The table shows the number of letters in the first 100 words of the novel *Pride and Prejudice* by Jane Austen.

| Word length | 1 | 2 | 3 | 4 | 5 | 6 | 7 | 8 | 9 | 10 | 11 | 12 | 13 |
|---|---|---|---|---|---|---|---|---|---|---|---|---|---|
| Frequency | 6 | 28 | 19 | 17 | 10 | 4 | 3 | 5 | 1 | 2 | 3 | 1 | 1 |

Find the mean length of a word.

**4** **a** Expand and simplify $(p + q)^2$.

    **b** Hence or otherwise, find the value of

       $1.28^2 + 2 \times 1.28 \times 2.72 + 2.72^2$

**5** Atul is carrying out a survey to find the most popular method of keeping fit. He asks 10 of his friends – all boys.

Explain why this sample could be biased.

**6** In a right-angled triangle, the difference between the other two angles is $12°$. Find the two missing angles.

**7** Solve

    **a** $x^2 + 7x + 10 = 0$        **b** $x^2 + 5x - 14 = 0$

    **c** $x^2 - 8x + 12 = 0$        **d** $x^2 - 10x - 17 = 0$

## 4    Revision homework

**1**   Calculate

   **a**   $727 \times 36$     **b**   $292 \times 51$     **c**   $329 \times 417$     **d**   $815 \times 146$

**2**   Using square grid paper, draw a set of axes.

   Label the $x$-axis from $-5$ to 8.

   Label the $y$-axis from 0 to 5.

   Draw a parallelogram with vertices $(1, 1)$, $(2, 2)$, $(5, 2)$, $(4, 1)$.

   Label the parallelogram $P$.

   **a**   Translate the parallelogram $P$ through $\begin{pmatrix} -5 \\ 1 \end{pmatrix}$.
       Label the image $Q$.

   **b**   Translate the parallelogram $P$ through $\begin{pmatrix} 2 \\ 2 \end{pmatrix}$.
       Label the image $R$.

**3**   **a**   Simplify $3a^2 b \times 8a^3 b^2$.

   **b**   Factorise completely $x^2 + xy - 2x - 2y$.

**4**   Calculate the value of the length marked by a letter.

   **a**
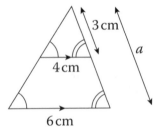

   **b**
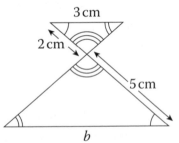

**5**   Write these fractions, decimals and percentages in ascending order.

   **a**   $\dfrac{4}{7}$, 0.23, 16%, 11%, 0.05, $\dfrac{11}{16}$, 25%, $\dfrac{2}{9}$

   **b**   0.75, $\dfrac{1}{13}$, 18%, $\dfrac{4}{5}$, 28%, 0.03, $\dfrac{4}{15}$, 0.4

**6**   Anna draws a rainbow.

   The outer arc is a semicircle of radius 8 cm.

   The inner arc is a semicircle of radius 6 cm.

   Calculate the area of the rainbow.

# Revision homework

**1** Given that $8.32 \times 640 = 5324.8$, write the answer to

  **a**  $83.2 \times 64$        **b**  $0.832 \times 6.4$       **c**  $532.48 \div 6.4$

  **d**  $0.832 \times 0.64$    **e**  $53.248 \div 83.2$    **f**  $5.3248 \div 0.64$

**2** **a** Work these out using mental methods.

    **i**  $\dfrac{3}{8}$ of 96      **ii**  15% of 240

    **iii**  $\dfrac{3}{7}$ of 105    **iv**  45% of 160

  **b** Write these fractions as percentages, using mental methods.

    **i**  $\dfrac{19}{25}$      **ii**  $\dfrac{17}{20}$      **iii**  $\dfrac{127}{200}$      **iv**  $\dfrac{23}{40}$

**3** Solve these pairs of simultaneous equations.

  **a**  $x + y = 5$      **b**  $4a + b = 7$      **c**  $2p - 3q = 7$

      $3x - y = 11$       $3a + 2b = 4$       $5p - 2q = 1$

**4** *AB* and *CD* are parallel.
Find the size of the angles marked *x* and *y*.
Give reasons for your answers.

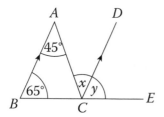

**5** The sector shown is folded to form a cone.
Find the curved surface area of the cone.

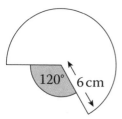

# Revision homework

**1  a  i**  Express 60 and 108 as products of their prime factors.

    **ii**  Use your answer to part **i** to work out the HCF of 60 and 108.

  **b**  If 1.42 × 230 = 326.6, write the answer to

    **i**   14.2 × 23      **ii**   0.142 × 2.3.

**2**  Work out each of these problems.

  **a**  Divide £30 in the ratio 1 : 4    **b**  Divide £120 in the ratio 5 : 7

  **c**  Divide 84 kg in the ratio 2 : 5    **d**  Divide 135p in the ratio 8 : 7

  **e**  Divide 105p in the ratio 5 : 2

**3**  Find the equation of each line on this graph.

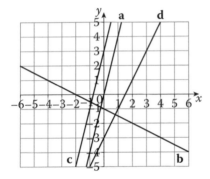

**4**  Isla and William are playing a computer game. They record the time (in seconds) that it takes to complete each round and compile the information shown in the table.

|  | Isla | Wiliam |
|---|---|---|
| **Median** | 64 | 59 |
| **Lower quartile** | 50 | 44 |
| **Upper quartile** | 71 | 73 |
| **Minimum** | 45 | 40 |
| **Maximum** | 75 | 82 |

  **a**  Draw box plots for each set of information on the same axes.

  **b**  Comment on and compare the performances of Isla and William.

**5**  350 g of organic tomatoes cost £ 1.61. Calculate the cost of 950 g of the same tomatoes.

# Revision homework

**1** Evaluate each of these

   **a** $4^{-\frac{1}{2}}$    **b** $5^0$    **c** $16^{\frac{3}{2}}$    **d** $125^{-\frac{2}{3}}$

**2** The first solution to each of these equations is given in brackets. Use a trigonometrical graph to find a second solution between $0°$ and $360°$.

   **a** $\sin x = 0.5$ $(x = 30°)$       **b** $\sin x = \dfrac{1}{\sqrt{2}}$ $(x = 45°)$

   **c** $\cos x = 0.5$ $(x = 60°)$       **d** $\cos x = \dfrac{\sqrt{3}}{2}$ $(x = 30°)$

**3** The cuboid and cone have the same surface area. Find the radius of the base of the cone.

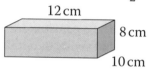
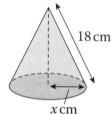

**4** Given that the equation of a line is $y = 3x + c$, work out the value of $c$ if the line passes through $(2, 1)$.

**5** The table gives information concerning the examination results of a group of 100 students.

   **a** Draw a cumulative frequency table and diagram for these data.

   **b** Estimate

      **i** the median

      **ii** the interquartile range from the graph.

   **c** Estimate the number of students that passed the examination if the pass mark was 55%.

| Test result, $t$% | Frequency |
|---|---|
| $40 < t \le 50$ | 5 |
| $50 < t \le 60$ | 20 |
| $60 < t \le 70$ | 34 |
| $70 < t \le 80$ | 27 |
| $80 < t \le 90$ | 12 |
| $90 < t \le 100$ | 2 |

**6** Phil measures the angle of elevation from the ground where he is lying to the top of a cliff as $23°$. He is exactly 60 m from the base of the cliff. By sketching a diagram and using trigonometry, work out the height of the cliff.

**7** Mac wants to invest £500 for 5 years. His bank offers him two options.

      Option 1 is simple interest of 4.25% per annum.

      Option 2 is compound interest of 4% per annum.

Which option should Mac choose in order to achieve the most interest on his investment? Show your working.

# 8    Revision homework

**1**   A health food shop sells five different types of dried fruit. Calculate the cost of 100 g of each type, given that

   **a**   200 g of apricots cost £2.20     **b**   600 g of figs cost £3.60

   **c**   350 g of blueberries cost £4.90   **d**   2 kg of raisins cost £8.00

   **e**   125 g of cranberries cost £2.00

**2**   Joe buys a new car for £15 995. A year later its value has depreciated by 18%. Work out its new value.

**3**   A bag contains 5 aquamarine, 3 amethyst and 2 topaz gemstones. A fair coin is thrown and a gemstone selected from the bag at random. Find the probability of obtaining

   **a**   a head and an aquamarine gemstone

   **b**   a tail and a topaz gemstone.

   **c**   What can you say about the events 'select an amethyst' and 'throw a head'? Explain your answer.

**4**   Calculate the unknown side $x$, leaving surds in your answer.

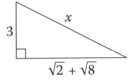

**5**   **a**   Copy and complete this table to draw the graph of $y = 2x - x^2 + 3$.

| $x$ | −2 | −1 | 0 | 1 | 2 | 3 | 4 |
|-----|----|----|---|---|---|---|---|
| $2x$ | −4 | | | | | | |
| $-x^2$ | −4 | | | | | | |
| $+3$ | +3 | | | | | | |
| $y$ | −5 | | | | | | |

   **b**   Draw the graph of $y = 2x - x^2 + 3$.

   **c**   Use your graph to solve

     **i**   $2x - x^2 + 3 = 0$       **ii**   $2x - x^2 + 3 = 2$.

   **d**   By drawing an appropriate graph on the same set of axes, find approximate solutions to $2x - x^2 + 3 = \frac{1}{2}x + \frac{3}{2}$.

   **e**   By drawing an appropriate graph on the same set of axes, find approximate solutions to $x - x^2 + 4 = 0$.

**6**   A cube has a surface area of $150\,\text{cm}^2$. Work out the volume of the cube.

# Revision homework

1 Find the *n*th term of these linear sequences.

    **a**  9, 11, 13, 15, 17, ...        **b**  17, 14, 11, 8, 5, ...

    **c**  $3\frac{1}{4}, 3\frac{1}{2}, 3\frac{3}{4}, 4, 4\frac{1}{4}, ...$

2 *y* is inversely proportional to *x*. If $y = 20$ when $x = 16$, find

    **a**  a formula for *y* in terms of *x*

    **b**  the value of *y* when $x = 10$

    **c**  the value of *x* when $y = 15$.

3 Write the equations of the three graphs labelled **a** to **c**.

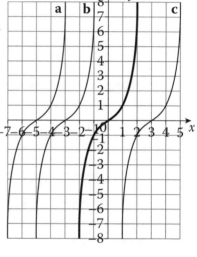

4 A metal worker melts down a solid metal sphere of radius 3 cm to form a cone of base radius 3 cm. Work out the height of the cone.

5 *PQRS* is a trapezium.

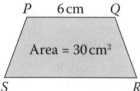

    *PQ* is 6 cm and $PQ : SR = 2 : 3$.
    The area of *PQRS* is 30 cm².
    Find the distance between *PQ* and *SR*.

6 A jacket cost £350.
    In the sale it cost £325.
    What was the percentage reduction?

7 *ABC* is a right-angled triangle.
    $CD : DB = 1 : 2$
    Angle $ABC = 43°$ and $AC = 14$ m.
    Calculate angle *CAD*.

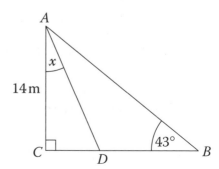

## 10    Revision homework

1. Julia bought her house for £250 000 in 2009. Two years later, an estate agent valued the house and deduced that the cost had risen by 34%. Work out the value of Julia's house in 2011.

2. A straight line has equation $y = \frac{1}{4}x - 2$.

   The point $P$ lies on the straight line. $P$ has a $y$-coordinate of $-1$.

   a   Find the $x$-coordinate of $P$.

   b   Write the equation of a straight line that is parallel to $y = \frac{1}{4}x - 2$.

3. Work out the surface area of this triangular prism which has a cross-section in the shape of an equilateral triangle of side 6 cm, height 5.2 cm and a length of 12 cm.

4. The table gives the marks of 10 students in the two papers of a French exam.

| Paper 1 | 75 | 64 | 50 | 80 | 45 | 58 | 72 | 63 | 50 | 74 |
|---------|----|----|----|----|----|----|----|----|----|----|
| Paper 2 | 81 | 70 | 52 | 89 | 42 | 62 | 72 | 73 | 48 | 79 |

   a   Represent these data on a scatter diagram.

   b   Describe the correlation shown in terms of the data.

   c   Draw a line of best fit and use this to estimate a mark in paper 1 for a student who was absent on the day of the exam but who scored a mark of 55 on paper 2.

5. Use the sine rule to find a solution for $\theta$. Use the sine graph to find a second solution. Sketch both solutions.

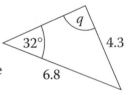

6. The curved surface area of a cylinder, radius $r$, is the same as the surface area of a sphere, radius $2r$. Show that the height of the cylinder is $8r$.

   Hint: Calculate the surface area of the sphere, radius $2r$, leaving the answer in terms of $\pi$ and $r$. Do the same for the curved surface area of the cylinder. Find $h$, the height of the cylinder.